Fundamental Data Compression

To my students, family and friends

Fundamental Data Compression

Ida Mengyi Pu

AMSTERDAM • BOSTON • HEIDELBERG • LONDON • NEW YORK • OXFORD
PARIS • SAN DIEGO • SAN FRANCISCO • SINGAPORE • SYDNEY • TOKYO
Butterworth-Heinemann is an imprint of Elsevier

Butterworth-Heinemann is an imprint of Elsevier
Linacre House, Jordan Hill, Oxford OX2 8DP
30 Corporate Drive, Suite 400, Burlington, MA 01803

First published 2006

British Library Cataloguing in Publication Data
A catalogue record for this book is available from the British Library

Library of Congress Cataloguing in Publication Data
A catalogue record for this book is available from the Library of Congress

ISBN–13: 978-0-7506-6310-6
ISBN–10: 0-7506-6310-3

For information on all Butterworth-Heinemann publications visit our web site at http://books.elsevier.com

Printed and bound in Great Britain

06 07 08 09 10 10 9 8 7 6 5 4 3 2 1

Contents

List of Figures

List of Algorithms

Preface

This book aims to introduce you to data compression, a fascinating research area in computer science as well as in telecommunication and computer networks of electronic engineering.

The book sets out a sequence for you to study the subject efficiently. It lays a good foundation for your professional work or advanced studies in the future. Topics include various useful compression techniques for text, audio, image, video data and international standards. I hope to show you the significance of some compression ideas and technologies, and, more importantly, to stimulate and encourage you to design your own compression algorithms.

We shall, in this text,

- discuss important issues in data compression
- study a variety of data compression techniques for compression of binary, text, sound and image data in digital forms
- learn how to design and implement compression algorithms.

We shall study some existing compression standards and compression utilities available. You will not only broaden the knowledge of compression techniques but also appreciate the mathematical impact on the evolution of the technology. You will also benefit from the development of your own transferable skills such as problem analysis and problem solving. Your skills in algorithm design and programming can also be improved by attempting the exercises, laboratory work and assessment questions at the end of each chapter.

I write this text book for students reading for their degrees as well as for anyone who is interested in data compression. Data compression is such a fruitful research area that the literature is extremely rich. However, I want to keep the contents of this book concise so the coverage and length of studies are suitable for a university single academic term (30 hours in 10 weeks) module. I want to give the reader many examples and illustrations to ease the learning process. I also want to include various exercises and implementation tasks to encourage readers' creative activities.

This book, therefore,

- looks at various topics in data compression from an algorithm designer's point of view

- is focused on the algorithmic issues which were overlooked in the past
- invites the reader to engage more in the teaching/learning process
- provides opportunities for the reader to apply the skills and knowledge from other modules, such as data structures, algorithm design, programming, software developments, internet computing, to name just a few.

Other considerations and arrangements include:

- following an order from simple to complex for sections and chapters
- covering certain theoretical foundations for data compression
- introducing not only the data compression techniques but also the ideas behind them
- focusing on algorithms and stimulating new ideas
- including exercises, laboratory problems, implementation hints, bibliography for further reading and assessment questions for each chapter.

Each data compression technique is viewed as a solution to a certain algorithmic problem in the book. The learning process can be viewed as a process of learning how to derive algorithmic solutions to various compression problems.

Bibliography

There are few textbooks on data compression for teaching purposes. However, a huge number of research papers and websites are dedicated to the subject.

To focus on the fundamentals, only a selective list is provided in the book for further reading. However, it should not be difficult for interested readers to find more detailed information on the Internet. Otherwise, the following key words can be used to start your search for various topics:

```
data compression
compression algorithm
information theory
run-length
Huffman coding
arithmetic coding
LZ77, LZ78, LZW
Burrows-Wheelers transform
```

Web page for the book

There will be an auxiliary website to the book (contact the publisher for details). It is a good idea to visit the page from time to time for updated information.

In addition to the usual teaching materials such as the texts or news, you may find our demonstration and experiment pages interesting. You may also check your understanding on certain concepts or verify your exercise results.

Get involved

The best way to learn a new subject is to get involved as much as possible. For example, you may like to share your own demonstration programme with others in the world. Check the book web page on how to contribute.

Prerequisites

This book is fairly self-contained. Appendices provide you with further mathematics background. However, there are some prerequisites including knowledge of elementary mathematics and basic algorithmics. You may find the issues in this book easier if you have reviewed certain topics in mathematics at undergraduate level, such as *sets, probability theory, basic computation on matrices* and *simple trigonometric functions* (e.g. $\sin(x)$ and $\cos(x)$, where $x \in [0, 1]$), and topics in *algorithm design*, such as *data structures* and *computational complexity*. It would be advantageous if you are fluent with one of computer programming languages.

For those who have neither the background nor the time to prepare themselves for the subject, we recommend that you follow each chapter closely since necessary mathematics or algorithmic foundations are discussed anyway.

Study methods

The highlight of each university academic year is to share the joy of my students' success in their studies. While different methods work better for different people, some methods seem to work for most people.

For example, one effective way to study compression algorithms is to trace the steps in each algorithm and attempt an example yourself. It is even better if you can follow the algorithmic ideas and try to invent your own.

Another effective way to study compression algorithms is to implement them and run experiments. Exercise and laboratory questions at the end of each chapter may direct you to various starting points for your experiments. They all together provide good help to your understanding.

Based on experience, we suggest and recommend the following practice:

1. Spend two hours on revision or exercise for every hour of study on new material.

2. Use examples to increase your understanding of new concepts, issues and problems.

3. Ask the question: 'Is there a better solution to the current problem?'

4. Use the Contents pages to comfort yourself with the scope of the subjects, and refer to the Learning Outcomes at the end of each chapter to clarify the learning tasks.

Experts have predicted that more and more people will be engaged in jobs involving certain multimedia application in the future. As more images and audio data are required to be processed, data compression techniques will continue to grow and evolve. Like any technology, what you have learnt today can become outdated tomorrow. We therefore recommend that you focus on the important principles of the subject and gain a good understanding of the issues in the field of data compression. The experience could be very useful for your future career.

Exercises, laboratory and assessment

The exercises, laboratory and assessment questions at the end of each chapter are set for you to check your understanding and to practise your programming skills.

It is useful for you to have access to a computer so you can implement the algorithmic ideas learnt from the book. There is no restriction on the computer platform nor a requirement for a specific procedural computer language. Our internal students at the University of London have gained experience in implementing various compression algorithms in *Java*, *C*, *C++*, *Python*, *Visual Basic*, *MatLab* or even *Pascal*.

Although implementation of an algorithm remains pretty much an art among university students today, you may like to follow a more systematic approach in your implementation in order to develop a 'good programming style':

1. Analyse and understand the algorithm.

2. Derive a general plan for the implementation.

3. Develop the program blocks.

4. Test or justify the correctness of the programs.

5. Comment on the limitations of the programs.

Your implementation details should be documented including a section on each of the above stages of work.

How to use this book

The book provides guidance for further development of your interests as well as further reading. However, you are not expected to read every item in the Bibliography section to enable individual topics to be studied in depth.

This book is written to invite you to get involved. The learning process requires the *input* of your own experiments and experience. Therefore, you are encouraged to, if possible, ask questions, pursue articles in research journals, browse the relative websites, attend conferences or trade shows etc., and in general pay attention to what is happening in the computing world. Best of all, try your own experiments and invent your own algorithms.

The rest of this book is organised as follows:

Chapter 1 discusses essentials, definitions and algorithmic concepts.

Chapter 2 introduces the information theory and enhances the concepts and issues in data compression.

Chapter 3 introduces an intuitive compression method: run-length coding. This simple algorithm serves as an example of how to design a compression algorithm in a systematic way.

Chapter 4 discusses the preliminaries of data compression and reviews the main idea of Huffman coding and Shannon-Fano coding. This serves as an example to demonstrate how to apply the information theory to analyse the compression algorithm, and how to address the efficient implementation issues.

Chapter 5 introduces adaptive Huffman coding.

Chapter 6 studies issues of arithmetic coding.

Chapter 7 covers dictionary-based compression techniques.

Chapter 8 introduces prediction and transforms. This serves as a foundation for the next three chapters.

Chapter 9 discusses one-dimensional wave signals. This serves as an application of prediction and transforms in the previous chapter.

Chapter 10 discusses image data and still image compression. This serves as an application of the prediction and transform techniques on two-dimensional data.

Chapter 11 introduces video compression methods.

Appendix A highlights the milestones in the area of data compression.

Appendix B reviews the basics on matrix operations.

Appendix C covers briefly the necessary mathematics for Fourier transforms.

Appendix D provides the guidance on the pseudocode in algorithm design.

And finally,

Appendix E gives a list of notations used in the book.

Your comments

If you have any comments about this book, either general or specific, favourable or unfavourable, you are very welcome to send them to `i.pu@gold.ac.uk`.

Good luck!

I.M. Pu
London
September 2005

Acknowledgements

I would like to take the opportunity to thank the many people who helped me in preparation of the book: David Hatter for initialising this project and encouragement; the reviewers Professors Maxime Crochemore, Bill Smyth, and an anonymous professional for valuable comments; the proof reader Dr Yuji Shen for helpful comments and suggestions; the editors and publishing team Alfred Waller, Melissa Read, Hayley Salter, Deb Puleston, and Jodi Cusack, for constant support.

I would also like to thank all the researchers working on original ideas of data compression for making this research area extremely interesting and challenging.

Chapter 1

Introduction

Data compression is, in the context of computer science, the science (and art) of representing information in a compact form. It has been one of the critical enabling technologies for the ongoing digital multimedia revolution for decades.

Most people frequently use data compression software such as `zip`, `gzip` and `WinZip` (and many others) to reduce the file size before storing or transferring it in media. Compression techniques are embedded in more and more software and data are often compressed without people knowing it.

Data compression has become a common requirement for most application software as well as an important and active research area in computer science. Without compression techniques, none of the ever-growing Internet, digital TV, mobile communication or increasing video communication techniques would have been practical developments.

Typical examples of application areas that are relevant to and motivated by data compression include

- personal communication systems such as facsimile, voice mail and telephony
- computer systems such as memory structures, disks and tapes
- mobile computing
- distributed computer systems
- computer networks, especially the Internet
- multimedia evolution, imaging, signal processing
- image archival and videoconferencing
- digital and satellite TV.

Practical problems have motivated various researches in data compression. Equally, research in data compression has also been based on or stimulated other new subject areas. Partly due to its broad application territory, data compression overlaps with many science branches and can be found in many different subject areas. For example, you will see chapters or sections dedicated to data compression in books on

- information theory
- coding theory
- computer networks and telecommunications
- digital signal processing
- image processing
- multimedia
- steganography
- computer security.

The language used in unrelated disciplines can be substantially different. In this book, the word *data* is in general used to mean the information in digital form on which computer programs operate, and *compression* means a process of removing redundancy in the data. By 'compressing data', we actually mean deriving techniques or, more specifically, designing efficient algorithms to:

- represent data in a less redundant fashion
- remove the redundancy in data
- implement compression algorithms, including both compression and decompression.

The interests and goals also tend to be diverse among people with different disciplinary backgrounds. This book focuses on the algorithmic aspects of data compression. We view data compression as a process of deriving algorithmic solutions to a compression problem. An algorithmic problem is a general question to be answered by an ordered sequence of instructions. The instruction sequence is regarded as a sequential algorithm for the problem as well as the solution to the problem. The algorithm allows a solution to any instance of the problem to be derived by execution of the algorithm. For example, a searching problem may be defined as follows:

Given a set s of elements and a target x, is the target x in the set?

This question is 'general' because it includes many instances. The set can contain any collection of elements and the target x can be any one of the same type. For instance, if $s = (12, 34, 2, 9, 7, 5)$, is $x = 7$ in the list? The algorithmic solution to this problem is to find an algorithm which derives an answer to every instance of the problem. A native algorithm would be the so-called *sequential search* as in Algorithm 1.1. The string is stored in a one-dimensional array $L[i]$.

The algorithm is written in *pseudocode* that is close enough to most conventional high-level computer languages. The advantage of using pseudocode is to allow the algorithm design to concentrate on the algorithmic ideas instead of being distracted by syntax details of a certain computer language. We shall present most algorithms in the book in pseudocode (see Appendix D).

Algorithm 1.1 Sequential search

INPUT: element list L and Target
OUTPUT: index where Target is found or 'not found'

read Target; set index i to 1
while $L[i] \neq$ Target and not end of the list **do**
    $i \leftarrow i + 1$
end while
if $L[i] =$ target **then**
    return i
else
    output 'not found'
end if

1.1 Data compression problems

A compression problem involves finding an efficient algorithm to remove various redundancy from a certain type of data. The general question to ask here would be, for example, given a string s, what is the alternative sequence of symbols which takes less storage space? The *solutions* to the compression problems would then be the compression algorithms that will derive an alternative sequence of symbols which contains *fewer* number of bits in total, plus the decompression algorithms to recover the original string.

How many fewer bits? It would depend on the algorithms but it would also depend on how much the redundancy can be extracted from the original data. Different data may require different techniques to identify the redundancy and to remove the redundancy in the data. Obviously, this makes the compression problems 'hard' to solve because the general question is difficult to answer easily when it contains too many instances. Fortunately, we can take certain constraints and heuristics into consideration when designing algorithms.

There is no 'one size fits all' solution for data compression problems. In data compression studies, we essentially need to analyse the characteristics of the data to be compressed and hope to deduce some patterns in order to achieve a compact representation. This gives rise to a variety of data modelling and representation techniques, which are at the heart of compression techniques.

1.1.1 Compression

Data compression can be viewed as a means for efficient representation of a digital source of data such as text, image, sound or any combination of all these types such as video. The goal of data compression is to represent a source in digital form with as few bits as possible while meeting the minimum requirement of reconstruction of the original.

In the context of this book, we regard data compression (or compression in short) as algorithms to achieve the compression goals on the source data.

Behind each algorithm there are ideas, mathematical models or implementation techniques to achieve the compression.

When working on compression problems, we need to consider the efficiency aspect of the algorithms as well as the effectiveness of compression. Intuitively, the behaviour of a compression algorithm would depend on the data and their internal structure. The more redundancy the source data has, the more effective a compression algorithm may be.

1.1.2 Decompression

Any compression algorithm will not work unless a means of decompression is also provided due to the nature of data compression. When compression algorithms are discussed in general, the word *compression* alone actually implies the context of both compression and decompression.

In this book, we sometimes do not even discuss the decompression algorithms when the decompression process is obvious or can be easily derived from the compression process. However, as a reader, you should always make sure that you know the decompression solutions as well as the ones for compression.

In many practical cases, the efficiency of the decompression algorithm is of more concern than that of the compression algorithm. For example, movies, photos, and audio data are often compressed once by the artist and then the same version of the compressed files is decompressed many times by millions of viewers or listeners.

Alternatively, the efficiency of the compression algorithm is sometimes more important. For example, the recording audio or video data from some real-time programs may need to be recorded directly to a limited computer storage, or transmitted to a remote destination through a narrow signal channel.

Depending on specific problems, we sometimes consider compression and decompression as two separate synchronous or asynchronous processes.

Figure 1.1 shows a platform based on the relationship between compression and decompression algorithms.

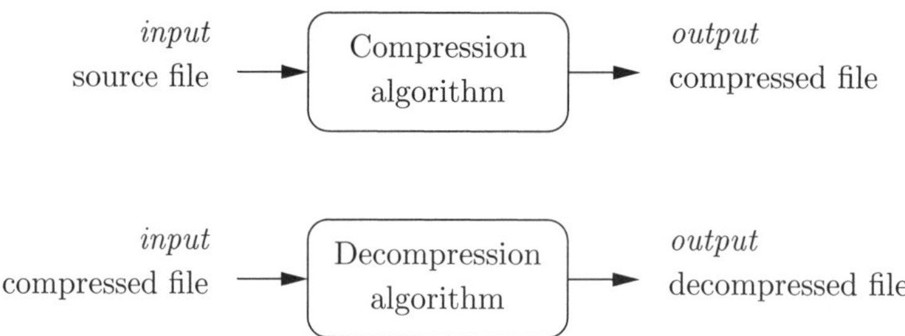

Figure 1.1: Compressor and decompressor

A compression algorithm is often called *compressor* and the decompression algorithm is called *decompressor*.

The compressor and decompressor can be located at two ends of a communication channel, at the *source* and at the *destination* respectively. In this case, the compressor at the source is often called the *coder* and the decompressor at the destination of the message is called the *decoder*. Figure 1.2 shows a platform based on the relationship between a coder and decoder connected by a transmission channel.

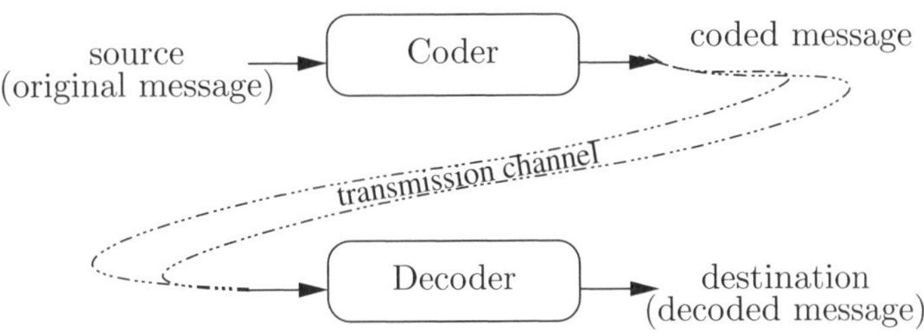

Figure 1.2: Coder and decoder

There is no substantial difference between the platform in Figure 1.1 and that in Figure 1.2 in terms of the compression algorithms discussed in this book. However, certain concepts may be discussed and understood more conveniently at one platform than the other. For example, it might be easier to introduce the information theory in Chapter 2 based on the coder-decoder platform. Then again, it might be more convenient to discuss the symmetric properties of a compression algorithm and decompression algorithm based on the compressor-decompressor platform.

1.2 Lossless and lossy compression

There are two major families of compression techniques when considering the possibility of reconstructing exactly the original source. They are called *lossless* and *lossy* compression.

Lossless compression

A compression approach is lossless only if it is possible to exactly reconstruct the original data from the compressed version. There is no loss of any information during the compression[1] process.

For example, in Figure 1.3, the input string `AABBBA` is reconstructed after the execution of the compression algorithm followed by the decompression algorithm.

Lossless compression is called *reversible* compression since the original data may be recovered perfectly by decompression.

[1]This general term should be read as both compression and decompression.

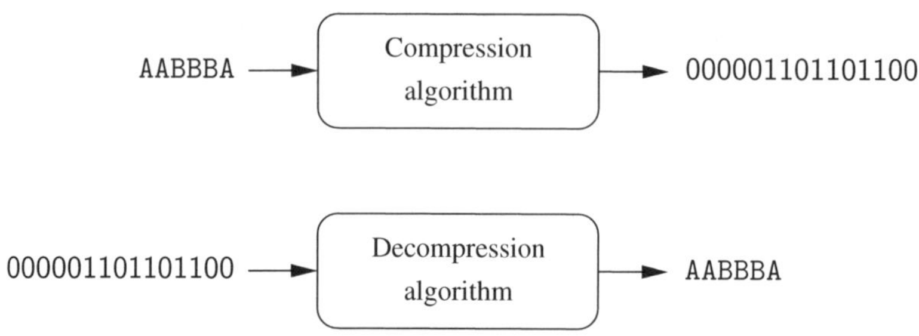

Figure 1.3: Lossless compression algorithms

Lossless compression techniques are used when the original data of a source are so important that we cannot afford to lose any details. Examples of such source data are medical images, text and images preserved for legal reason, some computer executable files, etc.

Lossy compression

A compression method is lossy if it is not possible to reconstruct the original exactly from the compressed version. There are some insignificant details that may get lost during the process of compression. The word insignificant here implies certain requirements to the quality of the reconstructed data.

Figure 1.4 shows an example where a long decimal number becomes a shorter approximation after the compression-decompression process.

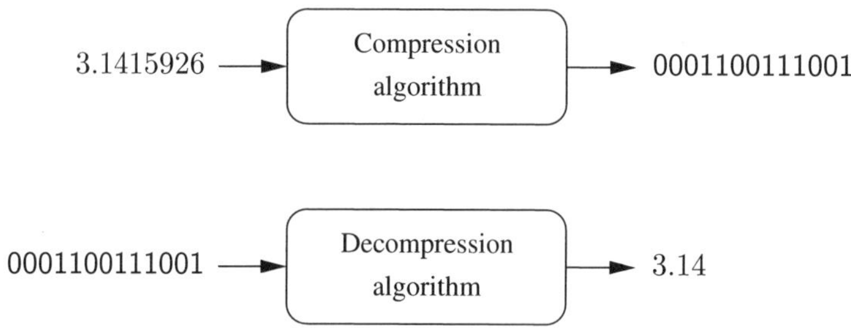

Figure 1.4: Lossy compression algorithms

Lossy compression is called *irreversible* compression since it is impossible to recover the original data exactly by decompression.

Approximate reconstruction may be desirable since it may lead to more effective compression. However, it often requires a good balance between the visual quality and the computation complexity.

Data such as multimedia images, video and audio are more easily compressed by lossy compression techniques because of the way that human visual and hearing systems work.

One important motivation for data compression is to allow some efficient transmission of the data via a transmission channel. It is convenient, for the rest of this section, to consider the coder-decoder platform that we mentioned in the previous section. Obviously, the amount of data was restricted by the capacity of the transmission media, which is called *bandwidth* and measured in *bits per second.*

Considering the effect of lossy compression, there are two kinds of classic compression problems of interest, namely

- **Distortion-rate problem** Given a constraint on transmitted data rate or storage capacity, the problem is to compress the source file at, or below, this rate but at the highest fidelity possible.

  Compression in areas of voicemail, digital cellular mobile radio and videoconferencing are examples of distortion-rate problems.

- **Rate-distortion problem** Given the requirement to achieve a certain pre-specified fidelity, the problem is to meet the requirements with as few bits per second as possible. Compression in areas of CD quality audio and motion picture quality video are examples of rate-distortion problems.

1.3 Deriving algorithmic solutions

There are many ways to design algorithms. A systematic approach including eight stages can be summarised as follows:

1. Description of the problem
2. Mathematical modelling
3. Design of the algorithm
4. Verification of the algorithm
5. Estimation of the computational complexity
6. Implementation
7. Program testing
8. Documentation.

Stage 1: Description of the problem

A compression problem, from the algorithmic point of view, is to find an effective and efficient algorithm to remove various redundancy from certain types of data.

In this stage, we want to understand the problem precisely and ask the right questions. For example, we want to know what the format of the data is and what the restrictions to the output might be. We need to understand the problem well enough to be able to write a precise statement of the problem. If the problem is vague, we may use methods such as *divide and conquer* to divide the problem into subproblems, and to divide the subproblems into their subproblems repeatedly until every subproblem is manageable.

The deliverables of this stage are a specification of the problem which should include details about the input and output of the algorithm. For example, we would decide if the problem is lossless or lossy in nature. For lossy compression, we would consider whether the problem can be classified as a distortion-rate or rate-distortion problem.

Stage 2: Mathematical modelling

Modelling is a process of setting up an environment to allow the interested variables to be observed or certain behaviour of a system to be explored. It is a formalisation and extension to the description of a problem. Rules and relationships are often set in mathematical formula.

Modelling is a critical stage of algorithm design. A model can sometimes decide immediately the approaches of the algorithm. As we can see from later chapters, a good model can lead to efficient algorithmic solutions. Certain preconditions are usually assumed and the environment is described in mathematical terms. In data compression, models are used to describe a source data.

For compression problems, modelling can be viewed as a process of identifying the redundant characteristics of the source data, and finding an effective way to describe them. The model is the embodiment of what the compression algorithm knows about the source. It is like a platform on which every compression algorithm has to make use of some knowledge in order to perform. For example, the Huffman algorithm is based on a probability distribution of the source alphabet.

Some problems, however, are not easy and even sometimes impossible to model. Researchers have constantly been looking for better models for various compression problems. Mathematical modelling is an important branch in mathematics, statistics and computer science as a popular research area in its own right. In data compression, as one of its application areas, commonly used mathematical models have been developed over the years. Hence in this stage, priority may be given to finding a good model instead of actually building a new one from scratch.

The commonly used models for data compression can be classified as follows:

1. Physical model: using known physics of the source such as certain data generation processes or empirical observation
2. Probability models: using probability theory to describe the source
3. Markov model: using Markov chain theory to model the source
4. Composite model: describing the source as a combination of several different kinds and using a switch to activate one type at a time.

The deliverables at the end of the modelling stage include a feasible model for the compression problem of interest, which accommodates the redundancy of the data and from which the output constraints are met under some well-defined relationship between the input and output of the algorithm.

We try to show you in this book why a specific model is good for certain compression problems and the difference a choice would make between two different models. You should note how a model is substantially influenced on the algorithmic solutions in later stages.

Stage 3: Design of the algorithm

In this stage, we may apply literally all the algorithm knowledge and techniques we have.

Design of the algorithms is an interesting and challenging task. The techniques depend highly upon the choice of the mathematical models. We may add further details to the model, consider feedback to realise the model, using standard techniques in algorithm design. For example, we may decide to use certain data structures, abstract data types or various off-the-shelf tools to help us with algorithmic solutions. We may take top-down approaches, and identify existing efficient algorithms to achieve partial solutions to the problem.

Most data compression problems are data oriented. There is unlikely to be an efficient algorithm to a general question about all sorts of data. We may then have to adjust the data range, or add more restrictions to the specific type of data. We may even return to the previous stage and experiment with alternative models if serious flaws are found in the approach.

The deliverables of this stage are the correct algorithmic solutions to our problem. This includes algorithms in pseudocode and convincing consideration on data structures.

In this book, we shall try to highlight some good practice of the algorithm design whenever possible and extend your experience on algorithm development.

Stage 4: Verification of the algorithm

This is sometimes the most difficult task. We may have seen or heard of a common practice in software development in which people tend to leave this until the program testing stage. The drawback in that approach is that we may have wasted an enormous amount of energy and time on programming before realising the algorithm has fundamental flaws.

In this stage, we check the correctness of the algorithm, the compression quality, and efficiency of the coder. It is relatively easy to check the compression quality. For example, we can use *compression ratio* or *saving percentage* to see how effective the compression is achieved. The coder efficiency is defined as the difference between the average length of the codewords and the entropy (see Chapter 2).

The deliverables at the end of this stage can be correctness proofs for the algorithmic solutions, or other correctness insurance such as reasoning or comparisons. Despite being difficult, the verification work in this stage is frequently proven to be extremely valuable.

In this book, we shall, whenever possible, try to offer some justification for algorithm design and show how the algorithms would achieve their objective output on *all* the appropriate input data.

Stage 5: Estimation of the computational complexity

Similar to algorithm verification, time and money have been regarded as well spent for analysing and estimating the efficiency of algorithms. A good analysis can save a lot of time and energy being wasted on faulty and disastrous software products.

In this stage, it is possible to estimate and predict the behaviours of the software to be developed using a minimum amount of resources, for example using just a pen and a piece of paper. We should therefore try to compare at least two candidate algorithms in terms of efficiency, usually time efficiency. The more efficient algorithm should also be checked to ensure that it has met certain theoretical bounds before being selected and implemented.

Stage 6: Implementation

Due to space limitations in this book, we leave all the implementation to the reader as laboratory exercises. However, guidance and hints are provided to encourage you to practise programming as much as you can.

There is no restriction on what hight-level computer language you use and how you would like to implement procedures or functions.

Stage 7: Program testing

This is a huge topic in its own right. There are formal methods that are dedicated to the work at this stage, but we shall not cover the details in this book.

Stage 8: Documentation

This is another important stage of work that this book has to leave to the reader.

However, you are encouraged to find an effective and consistent way to include this stage into your study activities. If you are too pressed for time, please try to add sufficient comments at least in your source programs.

From the discussion of the eight stages of the work above, it may have given us some ideas as to what we normally do in the whole process of algorithm design and software development. However, how would we evaluate what we do? How would we know that our compression algorithms are better than others, or vice versa?

We shall in the next section introduce the criteria and techniques commonly used to measure a compression algorithm.

1.4 Measure of compression quality

The performance of a compression algorithm can be measured by various criteria depending on the nature of the application. When time efficiency is not an issue (though it is equally important!), our main concern would be space efficiency, i.e. how effectively a data compression algorithm can save storage space. For example, a measure of percentage of difference in size of the input file before compression and the size of the output after compression would give a good indication of the effectiveness of the compression.

It is difficult to measure the performance of a compression algorithm in general because its compression behaviour depends greatly on whether data contain the redundancy that the algorithm looks for. The compression behaviour also depends on whether we allow the reconstructed data to be identical to the source data. We therefore shall discuss the measure in two situations, namely lossless compression and lossy compression.

Lossless compression

For lossless compression algorithms, we measure the compression effect by the amount of shrinkage of the source file in comparison to the size of the compressed version. Following this idea, several approaches can be easily understood by the definitions below:

- **Compression ratio** This is simply the ratio of the output to the input file size of a compression algorithm, i.e. the compressed file size after the compression to the source file size before the compression.

$$\text{Compression ratio} = \frac{\text{size after compression}}{\text{size before compression}}$$

- **Compression factor** This is the reverse of *compression ratio*.

$$\text{Compression factor} = \frac{\text{size before compression}}{\text{size after compression}}$$

- **Saving percentage** This shows the shrinkage as a percentage.

$$\text{Saving percentage} = \frac{\text{size before compression} - \text{size after compression}}{\text{size before compression}}\%$$

Note: some books (e.g.[Say00]) define the compression ratio as the compression factor defined here. The following example shows how the above measures can be used.

Example 1.1 *A source image file (256×256 pixels) with 65 536 bytes is compressed into a file with 16 384 bytes.*

Applying the definition above, we can easily work out that in this case the compression ratio is $16\,384/65\,536 = 1/4$, the compression factor is 4, and the saving percentage is 75%.

Note the performance of a compression algorithm cannot of course be reflected by one such instance. In practice, we may want to have a sequence of such tests to compute the average performance on a specific type of data, on text data only, for example. As we may see in later chapters, we discuss compression algorithms on certain types of data but hardly all types at the same time.

In addition, the effectiveness of a compression algorithm is only one aspect of the measure of the algorithm. In fact, the following criteria should normally be of concern to the programmers:

- **Computational complexity** This can be adopted from well-established algorithm analysis techniques. We may, for example, use the O-notation [CLRS01] for the time efficiency and storage requirement. However, compression algorithms' behaviour can be very inconsistent. Nevertheless, it is possible to use past empirical results.
- **Compression time** We normally consider the time for encoding and decoding separately. In some applications, decoding time is more important than encoding time. In other applications, they are equally important.
- **Entropy** If the compression algorithm is based on statistical results, then entropy (Chapter 2) can be used as a theoretical bound to the source to help make a useful quantity judgement. It also provides a theoretical guidance as to how much compression can be achieved.
- **Redundancy** In certain areas of compression, the difference between the average code length (Chapter 2) and the entropy of the source can be regarded as redundancy. In some other areas, the difference between a normal and uniform probability distribution is identified as redundancy. The larger the gap, the greater amount of the redundancy in the code. When the gap is zero, the code is said to be *optimal*.
- **Kolmogorov complexity** This measurement works better for theoretical proof than for practical settings. The complexity of the source data in a file can be measured by the length of the shortest program for generating the data.
- **Empirical testing** Measuring the performance of a compression scheme is difficult if not impossible. There is perhaps no better way than simply testing the performance of a compression algorithm by implementing the algorithm and running the programs with sufficiently rich test data. Canterbury Corpus provides a good testbed for testing compression programs. See `http://corpus.canterbury.ac.nz` for details.
- **Overhead** This measure is used often by the information technology industry. Overhead is the amount of extra data added to the compressed version of the data for decompression later. Overhead can sometimes be

large although it should be much smaller than the space saved by compression.

Lossy compression

For lossy compression, we need to measure the quality of the decompressed data as well as the compression effect. The word *fidelity* is often used to describe the closeness between the source and decompressed file. The difference between the two, i.e. the source before compression and the file after decompression, is called *distortion*. Often *approximate distortion* is used in practice. We shall look at lossy compression performance more closely later.

1.5 Limits on lossless compression

How far can we go with lossless compression?

Can we hope to find a universal compression algorithm at all? By universal, we mean an algorithm that can take any data file as an input and generate an output of smaller size, and that the original file can be exactly reconstructed by a decompression algorithm.

If not, what is the proportion of the files that a weak lossless compression algorithm, no matter how less effective, can achieve at best?

If the compression is not effective for the first time, can we compress the *compressed file* the second time, or repeat the compression process a number of times to achieve a larger overall compression percentage?

The following two statements may surprise you:

1. There is no algorithm that can compress all the files even by 1 byte.
2. There are only less than 1% of all the files that can be compressed losslessly by 1 byte.

Both statements are *true*, unfortunately. An informal reasoning for these statements is as follows:

1. If statement 1 were not true, i.e. suppose there were an algorithm that could compress any file given, we would have been able to use such a lossless compression algorithm to repeatedly compress a given file.

   Consider a large source file called `big.file` and a lossless compression program called `cmpres`, which is said to be able to compress any file. This includes the compressed file that was compressed already once by `cmpres`. Now this compressed file should again be input to the program `cmpres` and be compressed effectively, and output another compressed file. This means that we could then effectively repeat the compression process to a source file.

   By 'effectively', we mean that the compression ratio is always less than 1 everytime after running the program `cmpres`. In other words, the size of the compressed file is reduced every time after running the program cmpres. Consequently, cmpres(cmpres(cmpres($\cdots$cmpres(big.file)$\cdots$))), the output file after compression repeatedly a sufficient number of times, would be of size 0.

Now we have a contradiction because it would be impossible to losslessly reconstruct the original. Therefore, statement 1 is true.

2. To prove statement 2 to be true, we only need to

   (a) compute the proportion of the number of files that can be compressed by 1 byte shorter over the total number of possible files, and

   (b) show this proportion is less than 1%.

Compressing a source file can be viewed as mapping the file to a compressed version which is a different file. A file of n bytes long being compressed by 1 byte is therefore equivalent to mapping the file of n bytes to a file of $n-1$ bytes.

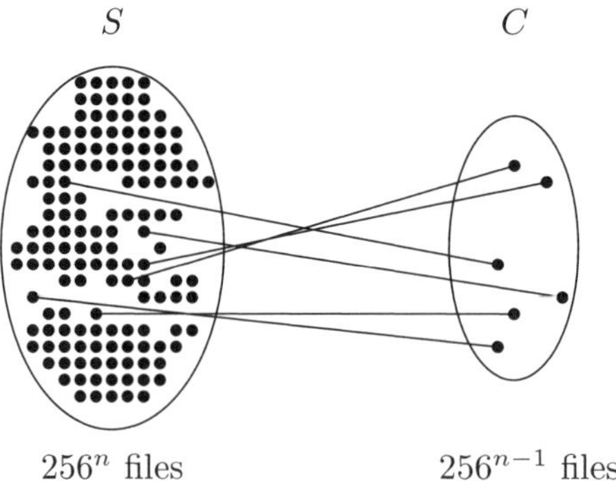

Figure 1.5: Mapping source files to their compressed version

In Figure 1.5, S on the left represents the set of all the files of n bytes long, and C on the right represents the set of all the files of $n-1$ bytes long. Each black dot in the set represents a unique file. Each line represents a mapping from a file of n bytes long to a file of $n-1$ bytes long.

There are $(2^8)^n = 256^n$ files of n bytes and 256^{n-1} of $n-1$ bytes in total. At best, every n byte long file can be compressed to an $n-1$ byte long file. Hence the number of the mappings is the number of files being compressed which is 256^{n-1} at most.

This means that the proportion of the successful one-to-one mappings is only $256^{n-1}/256^n = 1/256$. Clearly, this is less then 1% since $1/256 < 1/100 = 1\%$.

Therefore, only less than 1% of all the files can be compressed losslessly by 1 byte. In other words, no algorithm can actually compress 1% of all (possible) files even by 1 byte.

In fact, you can use a similar approach to find out more facts about how much a proportion of all files can or cannot be compressed.

Summary

Data compression is an interesting and active research area. There are many good reasons to study compression algorithms. Compression algorithms can be classified as two broad classes: lossless and lossy compressions. Our approaches are mainly algorithmic. Compression quality can be measured in various ways. The performance of lossless compression algorithms have limits. It can be shown that only a proportion of all the files can possibly be compressed losslessly.

Learning outcomes

On completion of your studies in this chapter, you should be able to:

- explain how to distinguish lossless data compression from lossy data compression
- outline the main compression approaches
- be aware of the main stages in the process of developing a compression algorithm
- estimate the effect and efficiency of a data compression algorithm
- explain the limits of lossless compression.

Exercises

E1.1 Investigate what compression software is available on your computer system. For example, do you use `winzip` or `gzip`, etc.?

E1.2 Suppose that you have compressed a file myfile using a compression utility available on your computer. What is the extension name of the compressed file?

E1.3 Use a compression facility on your computer system to compress a text file called myfile containing the following text:

This is a test.

Suppose you get a compressed file called myfile.gz after compression. How would you measure the size of myfile and myfile.gz?

E1.4 Suppose the size of myfile.gz is 20 KB while the original file myfile is 40 KB. Compute the compression ratio, compression factor and saving percentage.

E1.5 What is the main difference between an algorithm and a computer program?

E1.6 A compression process is often said to be 'negative' if its compression ratio is greater than 1.

Explain why negative compression is an inevitable consequence of a lossless compression.

E1.7 Analyse the statements below and give reasons why you are for or against them:

(a) Half of all files cannot be compressed by more than 1 bit.
(b) Three quarters of all files cannot be compressed by more than 2 bits.
(c) Seven-eighths of all files cannot be compressed by more than 3 bits.

Laboratory

L1.1 Design and implement a program that displays a set of English letters in a given string (upper case only).

For example, if the user types in a string '`AAABBEECEDE`', your program should display '`(A, B, E, C, D)`'.

The user interface should be something like this:

```
Please input a string:
> AAABBEECEDE
The letter set is:
(A, B, E, C, D)
```

L1.2 Write a method that takes a string (upper case only) as a parameter and returns a histogram of the letters in the string. The ith element of the histogram should contain the number of the ith character in the string alphabet.

For example, if the user types in a string '`AAABBEECEDEDEDDDE`', then the string alphabet is '`(A, B, E, C, D)`'. The output could be something like this:

```
Please input a string:
> AAABBEECEDEDEDDDE
The histogram is:
A xxx
B xx
E xxxxxx
C x
D xxxxx
```

L1.3 If you have access to a computer using the Unix or Linux operating system, can you use the `compress` or `gzip` command to compress a file?

L1.4 If you have access to a PC with Windows, can you use `WinZip` to compress a file?

L1.5 How would you recover your original file from a compressed file?

L1.6 Can you use the `uncompress` or `gunzip` command to recover the original file?

L1.7 Implement a program method `compressionRatio` which takes two integer arguments `sizeBeforeCompression` and `sizeAfterCompression` and returns the compression ratio.

L1.8 Implement a method `savingPercentage`which takes two integer arguments `sizeBeforeCompression` and `sizeAfterCompression` and returns the saving percentage.

Assessment

S1.1 Explain briefly the meanings of *lossless* compression and *lossy* compression. For each type of compression, give an example of an application, explaining why it is appropriate.

S1.2 Explain why the following statements are considered to be true in describing the absolute limits on lossless compression.

- No algorithm can compress all files, even by 1 byte.
- No algorithm can compress even 1% of all files, by just 1 byte.

Bibliography

[CLRS01] T.H. Cormen, C.E. Leiserson, R.L. Rivest, and C. Stein. *Introduction to Algorithms*. The MIT Press, 2nd edition, 2001.

[NG96] M. Nelson and J. Gailly. *The Data Compression Book*. M&T Books, New York, 2nd edition, 1996.

[Say00] K. Sayood. *Introduction to Data Compression*. Morgan Kaufmann, 2000.

Chapter 2

Coding symbolic data

Data compression is the science (and art) of representing information in a compact form. However, what is information? How would information be represented in a 'normal' form, i.e. the form before any compression? What do we mean by source data? How would we know if there is any redundancy in a source?

To answer these questions, we need first to clarify the meaning of terms such as *information*, *data*, *codes* and *coding*, and study the basics of information theory. Some conclusions and techniques learnt from this chapter will be very useful for later chapters.

2.1 Information, data and codes

Information is something that adds to people's knowledge. It is whatever contributes to a reduction in uncertainty of the human mind or the state of a system. People feel the existence of information, see media that carry information and react according to certain information all the time.

Information is not visible without some medium being a carrier. Data are the *logical* media often carried by some *physical* media such as a CD or a communication channel. Hence data can be viewed as a basic form of some factual information. This should be distinguished from other contrasting forms of information such as text, graphics, sound and image. A large amount of data can then be organised and stored in short messages or long files.

For example, data '$-30°$C' carry the factual information 'it is cold'. The same piece of information can be delivered by text 'minus thirty centigrade' on paper or by a picture of a thermometer on a computer screen, or by oral warning. Without these media, neither the data nor the information would have been visible.

The word *data* in the context of data compression includes any digital form of factual information that can be processed by a computer program. The data before any compression process are called the *source data*, or the *source* for short.

Examples of factual information may be classified broadly as text, audio, image and video. Many application programs adopt the information type as their data file type for convenience. Hence data may also be classified as text, audio, image and video while the real digital data format consists of 0s and 1s in a binary format.

- **Text** data are usually represented by 8 bit extended ASCII code (or EBCDIC). They appear in files with the extension `.txt` or `.tex` (or other coding system readable files like `.doc`) using an editor. Examples of typical text files are manuscripts, programs in various high-level languages (called source codes) and text emails.
- **Binary** data include database files, spreadsheet data, executable files, and program codes. These files usually have the extension `.bin`.
- **Image** data are represented often by a two-dimensional array of *pixels* in which each pixel is associated with its colour code. Extension `.bmp` represents a type of bitmap image file in Windows, and `.psd` for Adobe Photoshop's native file format.
- **Graphics** data are in the form of vectors or mathematical equations. An example of the data format is `.png` which stands for Portable Network Graphics.
- **Sound** data are represented by a wave (periodic) function. A common example is sound files in `.wav` format.

Three basic types of source data in the computer are *text*, (digital) *image* and *sound*. In application domains, the source data to be compressed are likely to be so-called *multimedia* and can be a mixture of *static media* format such as text, image and graphics, and *dynamic media* such as sound and video. Figure 2.1 demonstrates the stages involved for the source data in a file of a certain type to be encoded in a source binary file before compression. Figure 2.2 shows the reverse process in which the reconstructed binary data after decompression have to be decoded to data of a certain type before being recognised in any application.

2.2 Symbolic data

In this book, we often use the term *symbol*, or *character*, to mean a symbolic representation of input data to a compression algorithm. This is purely for the convenience of our algorithmic discussion. Under this notation, a symbol can be an audio sample, or an image pixel value as well as a letter, special character or a group of letters in a text. A text, image, audio or video file can then be considered as a one-dimensional, or multi-dimensional sequence of symbols. Figure 2.3 shows how the use of symbolic data would simplify the discussion of compression and decompression process. The simplified structure allows us to focus on the development of the compression algorithms.

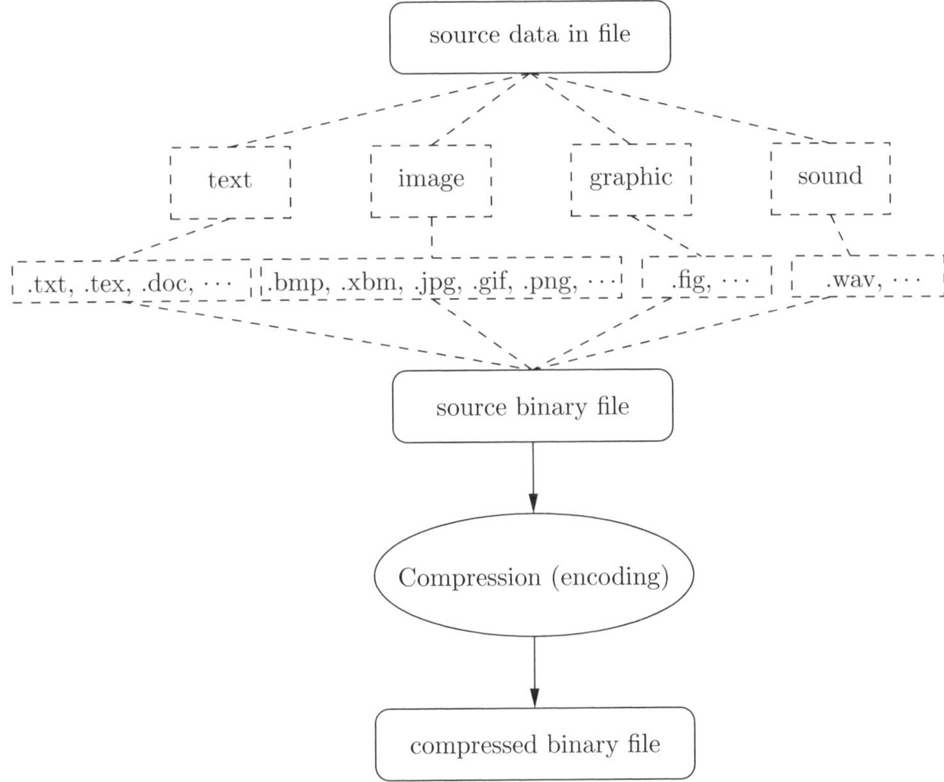

Figure 2.1: Data in compression

Suppose the alphabet of a source is $\mathcal{S} = (s_1, s_2, \cdots, s_n)$. The digital representation of the symbol set is called the *code* $\mathcal{C} = (c_1, c_2, \cdots, c_n)$ and the representation c_j of each symbol is called the *codeword* for symbol s_j, where $j = 1, 2, \cdots, n$. The process of assigning codewords to each symbol in a source is called *encoding*. The reverse process, i.e. to reconstruct the sequence of symbols in the source, is called *decoding*. Clearly, compression can be viewed as encoding and decompression as decoding in this sense.

The fundamental representation of data is ASCII code (pronounced 'ass-key') consisting of a set of *fixed length* (8 bit) codewords. It is possible to represent an alphabet by a set of *variable length* codewords and the code is then called a *variable length code*.

Example 2.1 *Two different binary codes,* $\mathcal{C}_1 = (000, 001, 010, 011, 100)$ *and* $\mathcal{C}_2 = (0, 100, 101, 110, 111)$*, can be used to represent the alphabet (*A*, B, C, D, E).*

Here $\mathcal{C}_1$ *is a fixed length code, as all the codewords consist of 3 bits.* $\mathcal{C}_2$ *is a variable length code since not all the codewords are of the same length.*

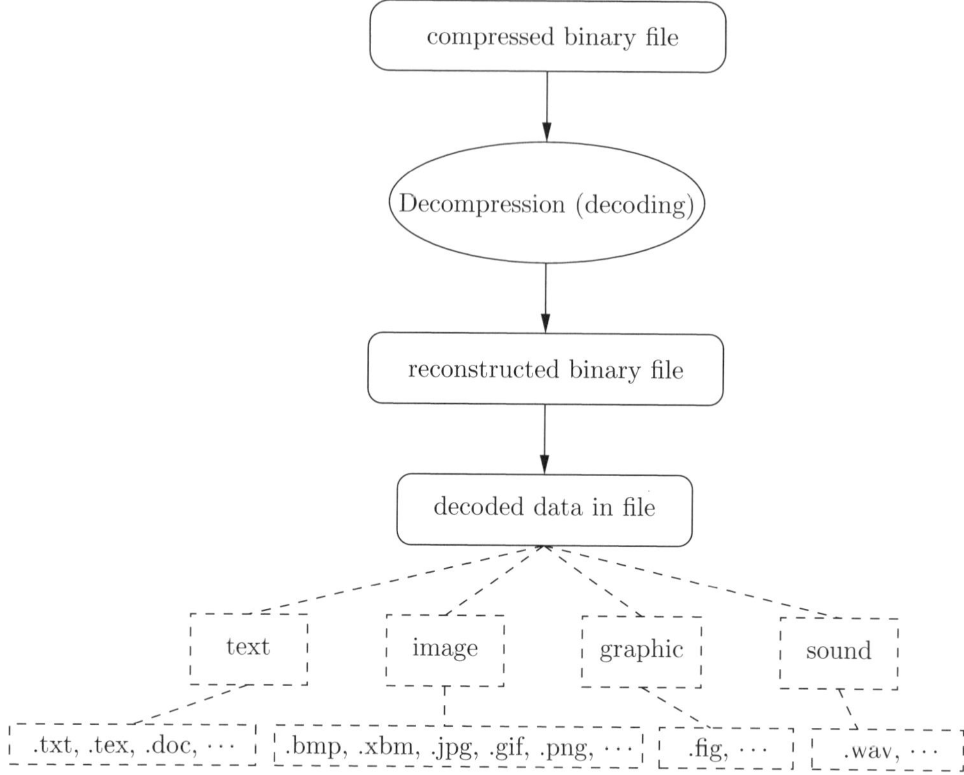

Figure 2.2: Data in decompression

It is important to distinguish the concept of *symbolic data*, *alphabet* and *code* in the context of a discussion. By symbolic data, we mean a source file consists of symbols from an alphabet. A code containing a set of codewords is usually a representation of the set of the alphabet.

For example, `BAAAAAAAC` are symbolic data from an alphabet such as (`A, B, C, D, E`). Suppose we define a fixed length binary code (000, 001, 010, 011, 100). Codewords 000, 001, 010, 011, 100 are the binary representation of `A, B, C, D, E` respectively. The binary representation of the symbolic data is `001 000 000 000 000 000 000 000 010` (without spaces). It is the source data in binary representation that, as an instance, is to be input into a compression algorithm. It is the size of this binary source data file that is to be hopefully reduced by a compression algorithm.

This can be seen clearly from Figure 2.4.

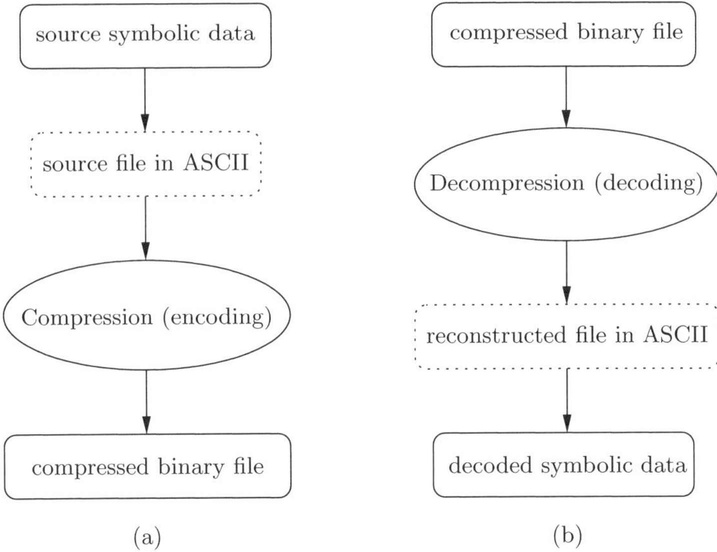

Figure 2.3: Symbolic data in compression

2.3 Variable length codes

Variable length codes are desirable for data compression because overall savings may be achieved by assigning short codewords to frequently occurring symbols and long codewords to rarely occurring ones.

For example, consider a variable length code (0, 100, 101, 110, 111) with lengths of codewords (1, 3, 3, 3, 3) for alphabet (A, B, C, D, E), and a source string `BAAAAAAAC` with frequencies for each symbol (7, 1, 1, 0, 0). The average number of bits required is

$$\bar{l} = \frac{1 \times 7 + 3 \times 1 + 3 \times 1}{9} \approx 1.4 \text{ bits/symbol}$$

This is almost a saving of half the number of bits compared to 3 bits/symbol using a 3 bit fixed length code.

The shorter the codewords, the shorter the total length of a source file. Hence the code would be a better one from the compression point of view.

2.3.1 Modelling

From the above example, we describe the problem of compression for symbolic data in a probability model below.

The source can be modelled by an alphabet $\mathcal{S} = (s_1, s_2, \cdots, s_n)$ and the probability distribution $\mathcal{P} = (p_1, p_2, \cdots, p_n)$ of the symbols.

Suppose we derive a code $\mathcal{C} = (c_1, c_2, \cdots, c_n)$ with length of each codeword $\mathcal{L} = (l_1, l_2, \cdots, l_n)$.

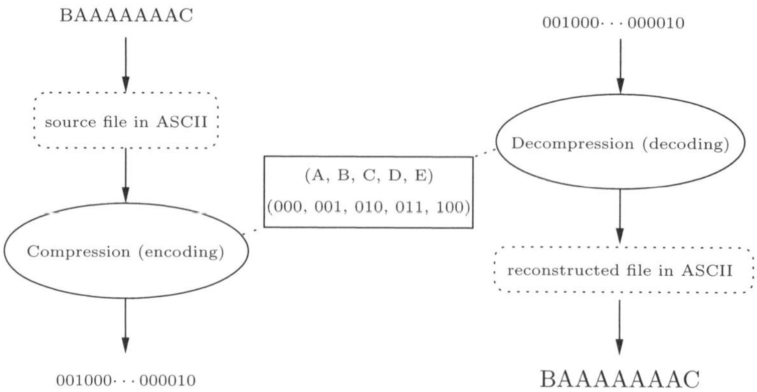

Figure 2.4: Code and source data

Our goal therefore is to minimise the *average length* of the code:

$$\bar{l}(\mathcal{P}, \mathcal{L}) = \sum_{j=1}^{n} p_j l_j$$

2.3.2 Unique decodability

Variable length codes are useful for data compression. However, a variable length code would be useless if the codewords could not be identified in a unique way from the encoded message.

Example 2.2 *Consider the variable length code (0, 10, 010, 101) for alphabet (A, B, C, D). A segment of encoded message such as '0100101010' can be decoded in more than one way. For example, '0100101010' can be interpreted in at least two ways, '0 10 010 101 0' as ABCDA or '010 0 101 010' as CADC.*

A code is *uniquely decodable* if there is only one possible way to decode encoded messages. The code (0, 10, 010, 101) in Example 2.2 is not uniquely decodable and therefore cannot be used for data compression.

Of course, we can always introduce an extra punctuation symbol during the encoding stage. For example, if we use symbol '/', we could then encode symbol sequence ABCDA as '0/10/010/101/0'. At the decoding end, the sequence '0/10/010/101/0' can be easily decoded uniquely. Unfortunately, the method is too costly because the extra symbol '/' has to be inserted for every codeword.

The ideal code in this situation would be a code not only of variable length but also with some *self-punctuating* property. For example, variable length code (0, 10, 110, 111) has such a self-punctuating property although the lengths for the codewords remain the same as those in (0, 10, 010, 101).

The self-punctuating property can be seen more clearly if we associate the codewords with the nodes of a binary tree in Figure 2.5. Each left branch is marked as 0 and the right branch as 1 in the binary tree. During decoding, each codeword can be obtained by collecting all the 0s and 1s from the root to each leaf. Every time a leaf is reached, we know that is the end of a codeword.

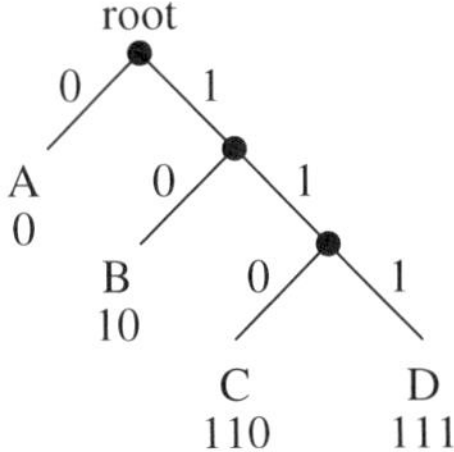

Figure 2.5: A prefix code

2.3.3 Prefix codes and binary trees

Codes with the self-punctuating property do exist. A type of so-called *prefix code* can be identified by checking its so-called *prefix-free property* or *prefix property* for short.

A *prefix* is the first few consecutive bits of a codeword. When two codewords are of different lengths, it is possible that the shorter codeword is identical to the first few bits of the longer codeword. In this case, the shorter codeword is said to be a prefix of the longer one.

Example 2.3 *Consider two binary codewords of different length:* $c_1 = 010$ *(3 bits) and* $c_2 = 01011$ *(5 bits).*

The shorter codeword c_1 is the prefix of the longer code c_2 as $c_2 = 01011$. Codeword c_2 can be obtained by appending two more bits 11 to c_1.

The prefix property of a binary code is the fact that *no* codeword is a prefix of another.

Example 2.4 *Consider the codewords in two codes (0, 10, 010, 101) and (0, 10, 110, 111).*

No codeword is a *prefix* of another in the uniquely decodable code (0, 10, 110, 111). In contrast, in code (0, 10, 010, 101), which is not uniquely decodable, codeword 0 is the prefix of codeword 010. Also codeword 10 is the prefix of codeword 101.

The prefix property turns out to be a favourite characteristic when searching for a uniquely decodable code. A code with such a prefix property is called a *prefix code*. In other words, a prefix code is a code in which no codeword is a prefix of another codeword, *neither* can a codeword be derived from another by appending more bits to a shorter codeword.

Example 2.5 *The code (1, 01, 001, 0000) is a prefix code since no codeword is a prefix of another codeword in the code. The code (0, 10, 110, 1011) is not a prefix code since 10 is a prefix of 1011.*

It is easy to check whether a binary code is a prefix code by drawing an associated binary tree. Each binary code can correspond to one such binary tree, in which each codeword corresponds to a path from the root to a node with the codeword name marked at the end of the path. Each bit 0 in a codeword corresponds to a left edge and each 1 to a right edge. Recall that, if a prefix code is represented in such an associate binary tree, all the codeword labels will be at its leaves (see Figure 2.5).

Two steps are involved in this approach:

1. **Construct the binary tree**

   First, we create a node as the root of the binary tree. Next, we look at the codewords one by one. For each codeword, we read one bit at a time from the first to the last. Starting from the root, we either draw a new branch or move down an edge along a branch according to the value of the bit.

   When a bit 0 is read, we draw, if there is no branch yet, a left branch and a new node at the end of the branch. We move down one edge along the left branch otherwise, and arrive at the node at the end of the edge. Similarly, when a bit 0 is read, we draw if there no branch yet, a right branch, or move down an edge along the right branch otherwise.

   The process repeats from node to node while reading the bit by bit until the end of the codeword. We mark the codeword after finishing with the whole codeword.

2. **Checking codeword position**

   If all the codeword labels are only associated with the leaves, then the codeword is a prefix code. Otherwise, it is not.

Example 2.6 *Decide whether the codes (1, 01, 001, 0000) and (0, 10, 110, 1011) for alphabet (A, B, C, D) are prefix codes.*

1. Draw a 0-1 tree as in Figure 2.6(a) and (b) for each code above.

2. For a prefix code, the codewords are only associated with the leaves. Since all the codewords in (1, 01, 001, 0000) are at leaves (Figure 2.6(a)), we can easily conclude that (1, 01, 001, 0000) is a prefix code.

   Since codeword 10 (B) is associated with an internal node of the 0-1 tree (Figure 2.6(b)), we conclude that (0, 10, 110, 1011) is not a prefix code.

Of course, for shorter codewords, we can easily draw the conclusion according to the definition of the prefix code. For example, noticing the second codeword 10 is the prefix of the last codeword 1011 in (0, 10, 110, 1011), we can easily decide that (0, 10, 110, 1011) is not a prefix code without drawing the binary tree.

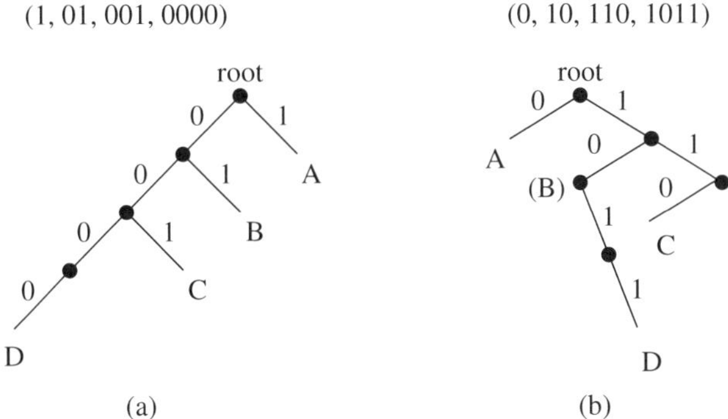

Figure 2.6: Prefix property and binary trees

2.3.4 Prefix codes and unique decodability

Prefix codes are a subset of the uniquely decodable codes. This means that all prefix codes are uniquely decodable. If a code is a prefix code, the code is then uniquely decodable.

However, if a code is not a prefix code, we cannot conclude that the code is not uniquely decodable. This is because other types of code may also be uniquely decodable.

Example 2.7 *Consider code (0, 01, 011, 0111) for (A, B, C, D). This is not a prefix code as the first codeword 0 is the prefix of the others, 01, 011 and 0111.*

However, given an encoded message 01011010111, there is no ambiguity and only one way to decode it: 01 011 01 0111, i.e. BCBD. Each 0 offers a means of self-punctuating in this example. We only need to watch out the 0, the beginning of each codeword and the bit 1 before any 0, the last bit of the codeword.

Some codes are uniquely decodable but require looking ahead during the decoding process. This makes them not as efficient as prefix codes.

Example 2.8 *Consider code (0, 01, 11) for (A, B, C). This is not a prefix code as the first codeword 0 is the prefix of the second codeword 01.*

Figure 2.7 shows a decoding process step by step to the encoded message 011101. The first 0 is read and it can be decoded as A. The next two 1s are decoded as C. However, an error occurs when the 0 is read after 1, because there is no codeword 10 in the code. This means the first choice was wrong: the first two bits 01 should have been decoded as B. Now the decoding process continue to decode the next two bits 11 as C. The following 0 can be decoded as A but only found an error again in the next step, because there is only one bit 1 left in the message which is not a codeword. Returning the last choice, the process reconsider 01 and decode them as B. Only by now, the decoding process is complete.

As we can see, the decoding process is not straightforward. It involves a 'trial and error' learning process and requires 'backtracking'. Twice the wrong choices of a codeword cannot be identified until a later stage and the message to be decoded needs to be reviewed repeatedly.

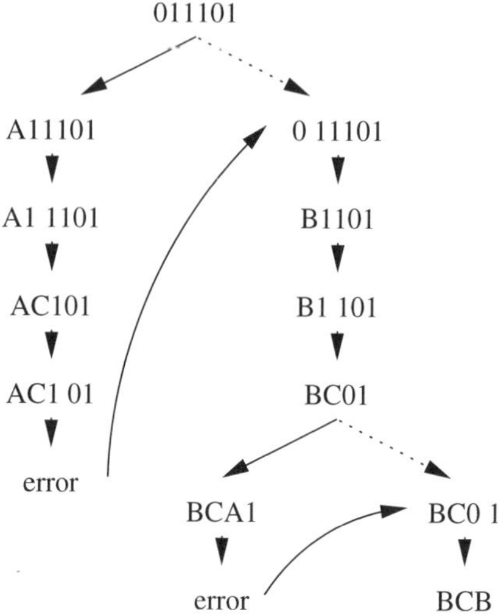

Figure 2.7: Not a prefix code but uniquely decodable

2.3.5 Kraft inequality

Prefix codes have another interesting property. For any non-prefix code whose codeword lengths satisfy certain conditions, we can always find a prefix code with the same codeword lengths.

Example 2.9 *Consider the uniquely decodable code (0, 01, 11) in Example 2.8. This is not a prefix code and the lengths of the codewords are 1, 2, 2 respectively.*

We can always find a prefix code, (0, 10, 11) for example, with the same codeword lengths.

The prefix property of a code guarantees only the correctness and efficiency of decoding. To achieve a good compression, the length of the codewords are required to be as short as possible.

Example 2.10 *Consider prefix code (0, 10, 110, 1111). The lengths of the codewords are 1, 2, 3 and 4. However, the length of the last codeword can be reduced from 4 to 3 as (0, 10, 110, 111) is also a prefix code with codeword lengths 1, 2, 3, 3. Figure 2.8(a) and (b) show the binary trees for code (0, 10, 110, 1111) and (0, 10, 110, 111) respectively. As we can see, if l_j is the length of the jth codeword, where $j = 1, \cdots, 4$, then the level at which the leaf for the codeword is $l_j + 1$.*

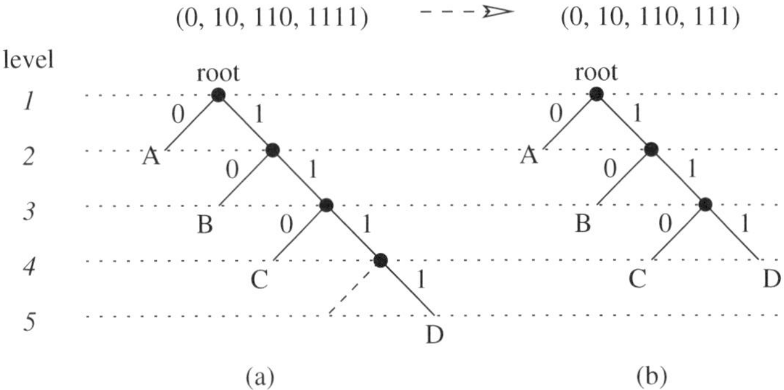

Figure 2.8: Two prefix codes

Can we reduce the codeword lengths further, even just by 1 bit? For example, would it be possible to find a prefix code with codeword lengths 1, 2, 3, 2 for symbols A, B, C and D respectively?

Example 2.11 *Discuss whether it is possible to find a prefix code with codeword lengths 1, 2, 3, 2.*

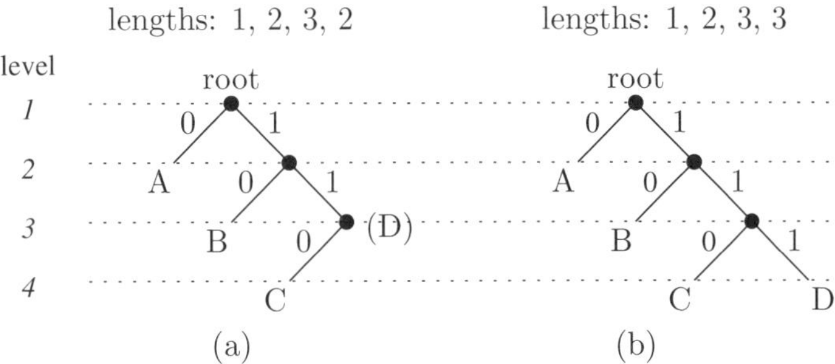

Figure 2.9: Impossible lengths (a) and possible lengths (b) for a prefix code

Solution We construct a binary tree according to the lengths given. Since it does not matter if a length 1 codeword is 0 or 1, we always build a branch for 0 first. We now have a codeword of length 1 (`0` labelled as A). Next we construct edges '1' and '0' and this gives a codeword of length 2 (`10` labelled as B). Next we add another '1' edge to get a codeword of length 3 (`110` labelled as C). Note in this approach constructing a codeword of length l means to find a path of length l from the root or to find a node at level l ($l = 2$ in this example, see Figure 2.9). There are three available paths of length 2 left: path 00, 01 and 11. Unfortunately, they are all contain an overlap with the paths corresponding to the codewords already constructed.

Suppose we choose to construct codeword `11` (labelled as D) and eventually derive a binary tree for a code in which codewords A, B, C, D are of lengths 1, 2, 3 and 2 respectively as in Figure 2.9(a).

As we can see from Figure 2.9(a), not all codewords are leaves. For example, symbol D is not at a leaf. This violates the fact that, for a prefix code, codewords should all be at the leaves in the binary tree. We therefore conclude that it is not possible to find a prefix code with codeword lengths 1, 2, 3, 2 for symbols A, B, C and D respectively.

For comparison, we also draw a binary tree for a code in which codewords are of lengths 1, 2, 3 and 3 as in Figure 2.9(b), where all the codewords are leaves.

Kraft's theorem provides a useful guide on the minimum requirements to the codeword lengths of prefix codes. Knowing the limit, we can avoid looking for a prefix code when it in fact does not exist.

Theorem 2.1 *There exists a prefix binary code* $\mathcal{C} = (c_1, c_2, \cdots, c_n)$ *with* n *codewords of lengths* $l_1, l_2, \cdots, l_n$ *respectively if and only if*

$$K(\mathcal{C}) = \sum_{j=1}^{n} 2^{-l_j} \leq 1$$

This inequality is known as the *Kraft inequality.*

With Kraft inequality, the question in Example 2.11 can be easily answered. If the lengths do *not* satisfy the Kraft inequality, we can conclude that it is not possible to find a prefix code consisting of these lengths.

Since in Example 2.11

$$K(\mathcal{C}) = \sum_{j=1}^{n} 2^{-l_j} = \frac{1}{2} + \frac{1}{2^2} + \frac{1}{2^3} + \frac{1}{2^2} > 1$$

it is impossible to find a prefix code with codeword lengths 1, 2, 3, 2.

Example 2.12 *Discuss the possibility of finding a prefix code with codeword lengths 1, 2, 3, 3.*

Solution Since

$$K(\mathcal{C}) = \sum_{j=1}^{n} 2^{-l_j} = \frac{1}{2} + \frac{1}{2^2} + \frac{1}{2^3} + \frac{1}{2^3} = 1$$

the lengths of the codewords satisfy the Kraft inequality, it is possible to find a prefix code with these codeword lengths (Figure 2.9(b)).

Observation

Kraft inequality can be misused if its claims are not carefully studied. We highlight here what the theorem *can* and *cannot* do.

1. The Kraft inequality sets requirements to the lengths of a prefix code. If the lengths do not satisfy the Kraft inequality, we know there is no chance of finding a prefix code with these lengths.

2. The Kraft inequality does not tell us how to construct a prefix code, nor the form of the code. Hence it is possible to find prefix codes in different forms and a prefix code can be transformed to another by swapping the position of 0s and 1s.

   For example, lengths (1, 2, 3, 3) satisfy Kraft inequality since

$$K(\mathcal{C}) = \sum_{j=1}^{n} 2^{-l_j} = \frac{1}{2} + \frac{1}{2^2} + \frac{1}{2^3} + \frac{1}{2^3} = 1$$

   Any one of the prefix codes (0, 10, 110, 111), (10, 0, 110, 111), (10, 110, 0, 111), (10, 110, 111, 0), (0, 110, 10, 111), (0, 110, 111, 10), (110, 0, 10, 111), (0, 10, 111, 110), $\cdots$ can be found.

   If (0, 10, 110, 111) is a prefix code, then (1, 01, 001, 000) is also a prefix code by just replacing 0s by 1s and 1s by 0s.

3. The Kraft inequality can tell that a given code is *not* a prefix code but it cannot be used to decide if a given code is a prefix code.

   When certain lengths satisfy Kraft inequality, it is possible to construct a prefix code in which the codewords are of these lengths. This implies there exist non-prefix codes with these lengths. Non-prefix codes can also be constructed with these lengths. Thus the code does not necessarily have to be a prefix code if its codeword lengths satisfy the Kraft inequality.

   For example, code (0, 01, 001, 010) satisfies the Kraft inequality since

$$K(\mathcal{C}) = \sum_{j=1}^{n} 2^{-l_j} = \frac{1}{2} + \frac{1}{2^2} + \frac{1}{2^3} + \frac{1}{2^3} = 1$$

   However, it is obviously not a prefix code because the first codeword 0 is the prefix of the others.

4. The Kraft inequality can tell us whether the lengths of a prefix code can be shortened, but it cannot make any change to the lengths.

   For example, consider the two codes in Example 2.10, (0, 10, 110, 1111) and (0, 10, 110, 111). The lengths of both codes satisfy the Kraft inequality. The lengths 1, 2, 3, 4 of the first code give

$$\frac{1}{2} + \frac{1}{2^2} + \frac{1}{2^3} + \frac{1}{2^4} < 1$$

   The lengths 1, 2, 3, 3 of the second code give

$$\frac{1}{2} + \frac{1}{2^2} + \frac{1}{2^3} + \frac{1}{2^3} = 1$$

   The Kraft inequality becomes equality when the code cannot be shortened.

Example 2.13 *Given an alphabet of four symbols (A, B, C, D), would it be possible to find a prefix code in which a codeword of length 2 is assigned to A, length 1 to B and C, and length 3 to D?*

Solution Here we have $l_1 = 2$, $l_2 = l_3 = 1$, and $l_4 = 3$.

$$\sum_{j=1}^{4} 2^{-l_j} = \frac{1}{2^2} + \frac{1}{2} + \frac{1}{2} + \frac{1}{2^3} > 1$$

Therefore, we cannot hope to find a prefix code in which the codewords are of these lengths.

Example 2.14 *If a code is a prefix code, what can we conclude about the lengths of the codewords?*

Solution Since prefix codes are uniquely decodable, they must satisfy the Kraft inequality.

Example 2.15 shows that, given a code with codeword lengths that satisfy the Kraft inequality, you cannot conclude that the code is a prefix code.

Example 2.15 *Consider code (0, 10, 110, 1011) for (A, B, C, D). This is not a prefix code as the second codeword 10 is the prefix of the last codeword 1011, despite the lengths of the codewords being 1, 2, 3, 4 which satisfy the Kraft inequality*

$$\frac{1}{2} + \frac{1}{2^2} + \frac{1}{2^3} + \frac{1}{2^4} < 1$$

However, since the lengths satisfy the Kraft inequality, we can always find a prefix code with the same codeword lengths (1, 2, 3, 4 respectively), such as (0, 10, 110, 1111).

2.4 Elementary information theory

Information theory is a study of information based on probability theory. It was proposed by Claude E. Shannon at Bell Laboratories in 1948 and based on people's intuition and reaction towards general information. It aims at a mathematical way of measuring the quantity of information.

As we mentioned in Section 2.1, information is something that adds to people's knowledge. The more a message conveys what is unknown (so it is new and surprising), the more informative the message is. In other words, the element of surprise, unexpectedness or unpredictability is relevant to the amount of information in the message.

For example, to most final year university students, the message 'Two thirds of the students passed exams in all subjects' offers less information than 'Two thirds of the students got 100% marks for exams in all subjects' although the two messages contain a similar number of words. The reason is the first message merely conveys something that happened frequently before and what one would normally expect, but the second message states something that would not normally happen and is totally unexpected.

The expectation of the outcome of an event can be measured by the *probability* of the event. The high expectation corresponds to a high probability of the event. A rarely happened event means a low probability event. Something that never happens has a zero probability. Hence the amount of information in a message can also be measured quantitatively according to the unexpectedness or surprise evoked by the event.

Suppose there are a set of n events $\mathcal{S} = (s_1, s_2, \cdots, s_n)$. $\mathcal{S}$ is called an *alphabet* if each s_j, where $j = 1, \cdots, n$, is a symbol used in a source message. Let the probability of occurrence of each event be p_j for event s_j. These probabilities of the events $\mathcal{P} = (p_1, p_2, \cdots, p_n)$ add up to 1, i.e. $\sum_{j=1}^{n} p_j = 1$. Suppose further that the source is *memoryless*, i.e. each symbol occurs independently with no dependence between successive events.

The amount of one's surprise evoked by the event is called the *self-information* associated with event s_j and is defined by a simple mathematical formula:

$$I(s_j) = \log_b \frac{1}{p_j}$$

or in negative logarithm

$$I(s_j) = -\log_b p_j$$

where the logarithm base (i.e. b in the formula) may be in:

- unit *bits*: called base two, $b = 2$
- unit *nats*: base e, $b = e$
- unit *hartleys*: base ten, $b = 10$.

The most commonly used logarithm for self-information is the base-two logarithm and hence its common unit is *bits*.

In this book, we use base two for logarithms, i.e. we use '$\log x$' to mean '$\log_2 x$' if not specified otherwise. So the self-information is

$$I(s_j) = -\log_2 p_j$$

The definition for self-information is convenient. It supports our intuition from daily experience about the information.

As we can clearly see from Figure 2.10, the amount of the self-information for event s_j is inversely proportional to the changes of its probability. When the probability of some event is 1, the self-information of the event is $\log(1) = 0$ since it offers no information. When the probability is close to 0, the amount of self-information, $\log \frac{1}{p_j}$ (or $-\log p_j$), is high. The amount of the self-information increases as p_j decreases from 1 to 0.

As we mentioned earlier, $\mathcal{S} = (s_1, s_2, \cdots, s_n)$ can be considered as an alphabet where each s_j is a symbol. The self-information can then be interpreted naturally as the amount of information each symbol s_j conveys.

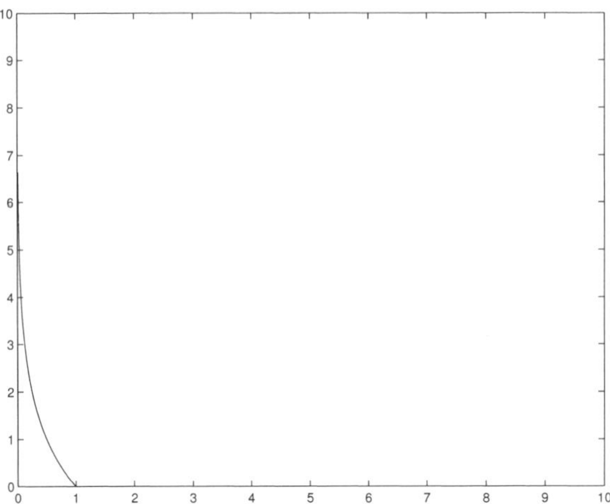

Figure 2.10: Self-information

Example 2.16 *If the probability of symbol 'd' occurring in a message is $\frac{1}{4}$, the self-information conveyed by the symbol is*

$$-\log\frac{1}{4} = -(-\log 2^2) = 2\log 2 = 2 \ bits$$

This suggests that it requires a binary codeword of minimum 2 bits for the information.

2.4.1 Entropy

Self-information is useful for measuring individual events or symbols at a time. However, we are often more interested in the information of a source where the occurrences of all possible symbols have to be considered. In other words, the measure of the information of a source has to be considered over the entire alphabet.

Example 2.17 *Consider a binary source $\mathcal{S} = (\mathtt{a}, \mathtt{b})$ with probabilities $\mathcal{P} = (3/4, 1/4)$ respectively. $I(\mathtt{a}) = -\log(3/4) = 0.415$ bits and $I(\mathtt{b}) = -\log(1/4) = 2$ bits.*

Here the self-information of each symbol in the alphabet cannot, on its individual basis, represent easily the whole picture of the source. One common way to consider a set of data is to look at its average value of the self-information of all the symbols. For example, the average self-information for each symbol in Example 2.17 is

$$\overline{I_1} = p_a I(\mathtt{a}) + p_b I(\mathtt{b}) = \frac{3}{4} \times 0.415 + \frac{1}{4} \times 2 = (1.245 + 2)/4 \approx 0.81 \text{ bit}$$

This can then be used to compare with another source.

Example 2.18 *Suppose we have another binary source $\mathcal{S}_2 = (\mathtt{c}, \mathtt{d})$ with probabilities $(\frac{1}{2}, \frac{1}{2})$ respectively. $I(\mathtt{c}) = I(\mathtt{d}) = -\log 1/2 = 1$ bit and*
$\overline{I_2} = p_c I(\mathtt{c}) + p_d I(\mathtt{d}) = 1$ bit.

We now see clearly that source $\mathcal{S}_2$ contains *on average* more information than source $\mathcal{S}_1$.

The average amount of the information, or the amount of the unexpectedness on average caused by a set of *independent* events , or over a set of *independent* symbols $\mathcal{S} = (s_1, s_2, \cdots, s_n)$, with probabilities of their occurrences $\mathcal{P} = (p_1, p_2, \cdots, p_n)$ respectively, is called the first-order *entropy* of the source or the *entropy* for short. It is defined as

$$H(\mathcal{P}) = \sum_{j=1}^{n} p_j I(s_j), \text{ or}$$

$$H(\mathcal{P}) = -\sum_{j=1}^{n} p_j \log p_j$$

Entropy[1] is the self-information of a set of symbols on average. It is the *mean* (or expected) amount of the self-information on $\mathcal{S}$, the event set or the alphabet. Note that the entropy of a source depends on the probability set $\mathcal{P} = (p_1, p_2, \cdots, p_n)$ instead of the values or the nature of $\mathcal{S} = (s_1, s_2, \cdots, s_n)$.

When $\mathcal{S} = (s_1, s_2, \cdots, s_n)$ is an alphabet, the entropy can be interpreted as a measure of, on average, the minimum number of binary symbols in bits that is required to encode the alphabet of the source.

Example 2.19 *Four symbols A, B, C and D occur with an equal probability in a source text file.*

Since $n = 4$ and $\sum_{j=1}^{4} p_j = 1$, the probabilities $p_1 = p_2 = p_3 = p_4 = \frac{1}{4}$. Therefore, the entropy of the source is

$$H(\mathcal{P}) = H(p_1, p_2, p_3, p_4) = \frac{1}{4}(-\log(\frac{1}{4})) \times 4 = 2 \text{ bits}$$

This can be interpreted as, on average, the source information requires a minimum of 2 bits to encode.

Example 2.20 *Suppose the same four symbols occur with probabilities 0.5, 0.25, 0.125 and 0.125 respectively in another source file.*

The entropy of the source is $H(\mathcal{P}) = H(0.5, 0.25, 0.125, 0.125) =$ $0.5 \times 1 + 0.25 \times 2 + 2 \times 0.125 \times 3 = 1.75$ bits.

This suggests that the average amount of information in the source is less than the source in the previous source. In other words, on average, the source information requires a minimum of 1.75 bits to encode.

[1] The term *entropy* was adopted by Shannon from thermodynamics where 'entropy' is used to indicate the amount of disorder in a physical system.

Example 2.21 *Suppose the probabilities are 1, 0, 0, 0. The entropy is:*

$$H(\mathcal{P}) = H(1, 0, 0, 0) = 0 \text{ bit}$$

This suggests that there is zero amount of information conveyed in the source. In other words, there is simply no need to encode the message.

Shannon showed that the best a lossless symbolic compression scheme using binary prefix codes can do is to encode a source with an average number of bits equal to the entropy of the source. This can be seen from the theorem below.

Theorem 2.2 *For any binary prefix code with the average codeword length* $\bar{l} = \sum_{j=1}^{n} p_j l_j$*, we have* $H(\mathcal{P}) \leq \bar{l} \leq H(\mathcal{P}) + 1$.

Proof To prove this is true, we only need to show that

1. The average codeword length is at least as long as the entropy: $H(\mathcal{P}) \leq \bar{l}$.

   This is true if and only if $H(\mathcal{P}) - \bar{l} \leq 0$. Following this idea, we write the difference

$$\begin{aligned} H(\mathcal{P}) - \bar{l} &= \sum_{j=1}^{n} p_j \log \frac{1}{p_j} - \sum_{j=1}^{n} p_j l_j \\ &= \sum_{j=1}^{n} p_j (\log \frac{1}{p_j} - l_j \log 2) \\ &= \sum_{j=1}^{n} p_j (\log \frac{1}{p_j} - \log 2^{l_j}) \\ &= \sum_{j=1}^{n} p_j (\log \frac{1}{p_j 2^{l_j}}) \end{aligned}$$

   Since $\ln x = \frac{\log x}{\log e}$, we have

$$\log \frac{1}{p_j 2^{l_j}} = (\ln \frac{1}{p_j 2^{l_j}}) \log e$$

   Therefore, the right side of the equation can be written

$$\begin{aligned} &= \sum_{j=1}^{n} p_j (\ln \frac{1}{p_j 2^{l_j}}) \log e \\ &= \log e \sum_{j=1}^{n} p_j (\ln \frac{1}{p_j 2^{l_j}}) \end{aligned}$$

Since $\ln x \leq x - 1$, for $x > 0$ the right side of the equation can be written

$$\begin{aligned}
&\leq \log e \sum_{j=1}^{n} p_j(\frac{1}{p_j 2^{l_j}} - 1) \\
&= \log e(\sum_{j=1}^{n} \frac{1}{2^{l_j}} - \sum_{j=1}^{n} p_j) \\
&= \log e(\sum_{j=1}^{n} 2^{-l_j} - 1)
\end{aligned}$$

According to the Kraft inequality, for a prefix code, we have

$$\sum_{j=1}^{n} 2^{-l_j} \leq 1$$

Therefore,

$$\sum_{j=1}^{n} 2^{-l_j} - 1 \leq 0$$

Multiplying this negative item by a positive $\log e$, the right side of the equation is again a negative, i.e.

$$H(\mathcal{P}) - \bar{l} = \log e(\sum_{j=1}^{n} 2^{-l_j} - 1) \leq 0$$

This proves that $H(\mathcal{P}) \leq \bar{l}$.

2. There exists a prefix binary code for which $\bar{l} < H(\mathcal{P}) + 1$.

   This can be justified by the following.

   As we know, the length of any binary code has to be an integer. If the prefix binary code is the optimum, then all its codeword length, for $j = 1, 2, \cdots, n$, $l_j = \log \frac{1}{p_j}$ when p_j are negative powers of 2, otherwise, $l_j > \log \frac{1}{p_j}$. This can be written as $l_j = \log \frac{1}{p_j} + \epsilon$, where $0 \leq \epsilon < 1$.

   Replacing ϵ by its exclusive upper bound on the right side of the equation, we have

$$l_j = \log \frac{1}{p_j} + \epsilon < \log \frac{1}{p_j} + 1$$

   Multiplying by $p_j > 0$ on both sides of the inequality, we have

$$p_j l_j < p_j(\log \frac{1}{p_j} + 1)$$

Therefore,

$$\begin{aligned}\bar{l} &= \sum_{j=1}^{n} p_j l_j \\ &< \sum_{j=1}^{n} p_j (\log \frac{1}{p_j} + 1) \\ &= \sum_{j=1}^{n} p_j \log \frac{1}{p_j} + \sum_{j=1}^{n} p_j \\ &= H(\mathcal{P}) + 1\end{aligned}$$

2.4.2 Optimum codes

Information theory can effectively tell how successful a prefix code is from a compression point of view. Given a prefix code, we know the length of each codeword $\mathcal{L} = (l_1, l_2, \cdots, l_n)$. If we know the source, i.e. $\mathcal{P} = (p_1, p_2, \cdots, p_n)$, we can compute the average length of the code $\bar{l} = \sum_{j=1}^{n} p_j l_j$, and the entropy of the source $H(\mathcal{P}) = -\sum_{j=1}^{n} p_j \log p_j$.

Since the entropy provides a theoretical bound of the minimum number of bits for the information, the difference between the average length of a code and the entropy represents the amount of the redundancy in the code.

The ratio in percentage of the entropy of a source to the average code length is a useful measure of the *code efficiency* for the source.

$$E(\mathcal{P}, \mathcal{L}) = \frac{H(\mathcal{P})}{\bar{l}(\mathcal{P}, \mathcal{L})} 100\%$$

Information theory says that the best a lossless symbolic compression scheme can do is to encode a source with an average number of bits equal to the entropy of the source. This is when $E(\mathcal{P}, \mathcal{L}) = 100\%$. The bigger the gap between the entropy and the average length of the code, the lower the $E(\mathcal{P}, \mathcal{L})$.

A code is *optimum* (or optimal) if the code efficiency reaches 100%. In other words, a code is optimal if the average length of the codewords equals the entropy of the source.

Example 2.22 *Consider a source alphabet $\mathcal{S} = (A, B, C, D)$ with probability distribution $\mathcal{P} = (\frac{1}{2}, \frac{1}{4}, \frac{1}{8}, \frac{1}{8})$. Suppose a prefix code $\mathcal{C} = (0, 10, 110, 111)$ is recommended. Comment on the efficiency of the code.*

Solution It is easy to check the code efficiency. We first write the lengths of the codewords $\mathcal{L} = (1, 2, 3, 3)$. The average length of the code is

$$\bar{l} = \sum_{j=1}^{4} p_j l_j = \frac{1}{2} \times 1 + \frac{1}{4} \times 2 + 2(\frac{1}{8} \times 3) = 1.75 \text{ bits}$$

The entropy of the source is

$$H = -\sum_{j=1}^{4} p_j \log p_j = \frac{1}{2} \times \log 2 + \frac{1}{4} \times \log 2^2 + 2(\frac{1}{8} \times \log 2^3) = 1.75 \text{ bits}$$

Therefore, the prefix code is optimal as the code efficiency is 100%.

2.5 Data compression in telecommunication

The concept of data compression in telecommunication can be easily understood from the diagram in Figure 2.11 on a message transformation system.

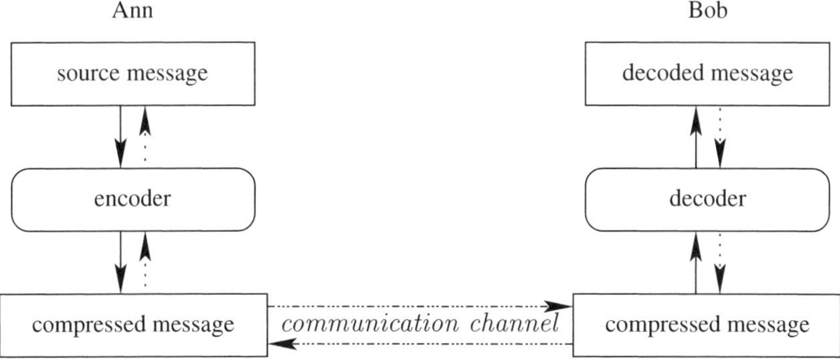

Figure 2.11: Communication

Suppose a piece of message is to be sent from one location by Ann and received at another location by Bob. The message is first encoded by an encoder before sending to the channel and the encoded message is decoded by a decoder on arrival at the other end and before being read by Bob.

What do the concepts such as *variable length codes*, *average length of codewords* and *entropy* mean in the telecommunication scenario?

Suppose that Ann and Bob are extremely good friends but they live far away from each other. They are both very poor students financially. Ann wants to send messages to Bob but has to save her money. Over the years, Ann and Bob have built their own secret alphabet. Suppose the next symbol that Ann wants to send to Bob is randomly chosen from an alphabet with a known probability distribution, and Bob knows the alphabet and the probability distribution.

Ann comes to ask for your advice with the important questions in the following example:

Example 2.23

1. *To minimise the average number of bits Ann uses to communicate her symbol to Bob, should she assign a fixed length code or a variable length code to the symbols?*

2. *What is the average number of bits needed for Ann to communicate her symbol to Bob?*
3. *What is meant by a 0 entropy? For example, what is meant if the probabilities associated with the alphabet are (0, 0, 1, $\cdots$, 0)?*

You give Ann the following solutions.

Solutions

1. Ann should use a variable length code because she is likely to use some symbols more frequently than others. Using variable length codes can hopefully save bits.
2. Ann needs at least the average number of bits that are equal to the entropy of the source. That is $-\sum_{i=1}^{n} p_i \log_2 p_i$ bits.
3. A '0 entropy' means that the minimum average number of bits that Ann needs to send to Bob is zero.

   Probability distribution (0, 0, 1, $\cdots$, 0) means that Ann will definitely send the third symbol in the alphabet as the next symbol to Bob and Bob knows this. If Bob knows what Ann is going to say then she does not need to say anything, does she?!

2.6 Redundancy

The first task in data compression is to identify any *redundancy* presented in the source data. Here the term redundancy has a general meaning. It can be some overlapped information, some common base data, some identical characteristics or some equivalent structures in nature, but all from saving storage point of view.

We shall try to identify various redundancies from the following simple examples:

Example 2.24 `BAAAAAAAC`, *a string that contains consecutive repeating characters.*

Here the redundancy is the 7 repeating symbols `A` which can be replaced by a shorter string such as r_7`A`.

Example 2.25 `ABACAA`, *a string that contains non-consecutive repeating characters.*

The redundancy in this example comes from the occurrence of symbol `A` under a fixed length code alphabet. There are more `A`s than any other symbols. If we use a shorter codeword to represent those more frequent symbols and a longer one for the less frequent ones, we may represent the string in a hopefully shorter form.

Example 2.26 *Consider a text with repeated words as follows:*

```
The red, the green and the blue colour, and
the paint in red, green or blue.
```

Here the redundancy is the repeated words which are patterns of strings such as `red`, `green` and `blue`.

Example 2.27 *Consider a vector of integers: (6, 428, 32, 67, 125).*

The redundancy in this example is the fact that the data cross over a big range [6, 428]. Consequently, each datum d in the vector requires 9 bits to represent since $0 < d < 512$ and $2^9 = 512$. This redundancy may be reduced by a simple scaling method. For example, applying d div 32 to the data, we have (0, 13, 1, 2, 4) of a much smaller range [0, 13]. Now only 4 bits are needed to represent each scaled datum d' as $0 \leq d' < 16$ and $2^4 = 16$.

Example 2.28 *Consider a matrix of binary data.*

```
000 001 011 011
001 001 001 010
011 001 010 000
```

Some binary patterns can be easily found as redundancy in this matrix. The two-dimensional data can be easily viewed as a sequence of string. For example, in `000 001 011 011 001 001 001 010 011 001 010 000` if we read the data one row after another.

However, hidden redundancy may become clearer if we divide the matrix into three 1-bit entry matrices as below:

1. Removing the last 2 bits of each entry:

```
0 0 0 0
0 0 0 0
0 0 0 0
```

2. Keeping the middle bit of each entry:

```
0 0 1 1
0 0 0 1
1 0 1 0
```

3. Removing the first 2 bits of each entry:

```
0 1 1 1
1 1 1 0
1 1 0 0
```

Figure 2.12: Pixels with similar colours

Now the first submatrix contains only consecutive 0s. The second one contains more consecutive 0s than 1s, and the last one contains more consecutive 1s than 0s.

Example 2.29 *Spacial redundancy.*

An image usually contains millions of pixels. Each pixel tends to be in the same or similar colour as its neighbours. Such a correlated relationship among neighbours is called *spatial redundancy.* Figure 2.12 shows a simple way to reduce the redundancy. The pixels on the left can be approximated to the one on the right as long as it is acceptable by the human visual system.

Example 2.30 *Quantisation.*

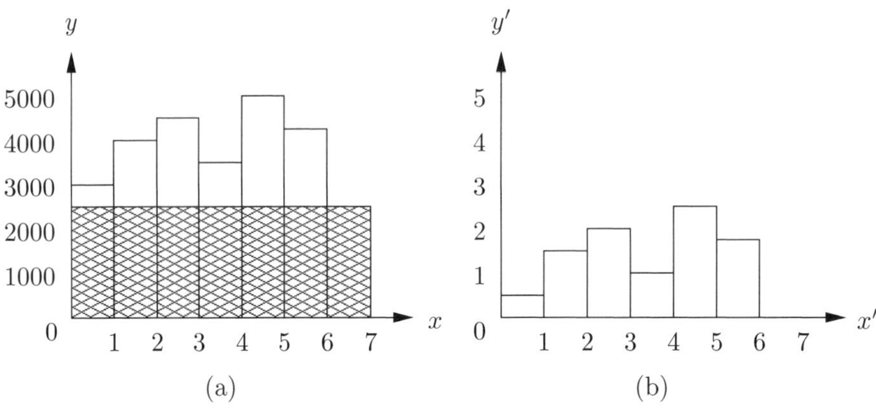

Figure 2.13: A histogram within certain ranges

In Figure 2.13(a), the shade area represents the redundancy. Let $x = x'$ and $y = 1000y' + 2500$ and we have a much simpler representation as in (b).

Example 2.31 *Transformation.*

In Figure 2.14, the redundancy can be seen from points on the left. The data are much easier to handle if we transform the coordinates of the points by rotating 32 degrees clockwise.

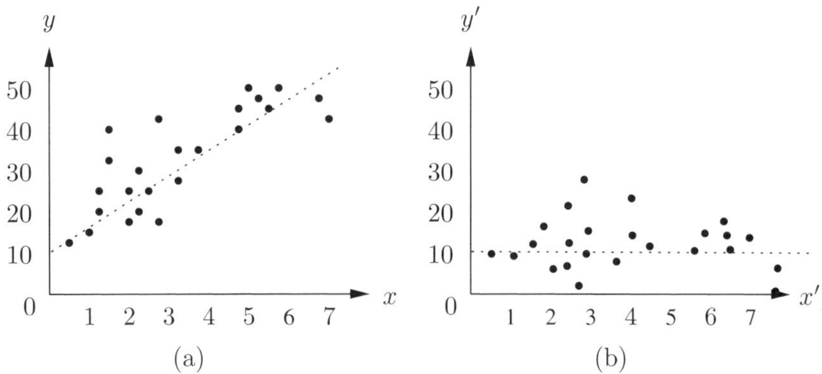

Figure 2.14: Some matrix data gathered along a line

Example 2.32 *Temple redundancy.*

Figure 2.15 shows a sequence of cartoon pictures (called *frames*), which is to be displayed one after another a few seconds apart. Due to the reaction of the human visual system, what we see would be a funny face changing from a sleep to awake, and finally to a 'smile' state.

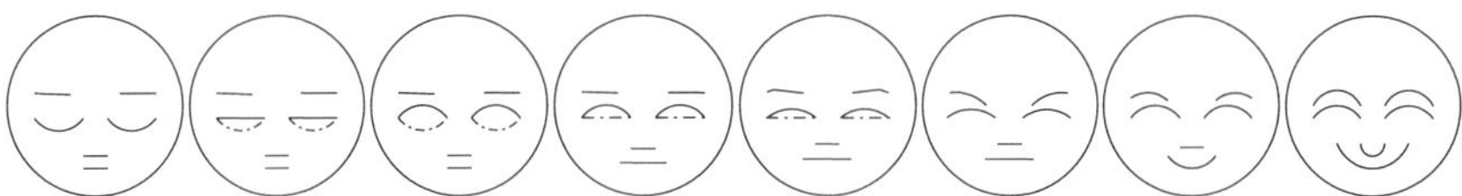

Figure 2.15: A sequence of cartoon frames

It is interesting to notice there is a large amount of redundancy here as the difference between one frame and the previous frame is tiny. For example, the first three faces are completely identical except the eyes. This kind of similarity is called *temple redundancy* and can be removed by simply storing the differences between frames.

We have tried to identify certain redundancy from the previous examples. As we can see, each example above shows a different kind of redundancy. Some redundancy can be viewed in more than one way and this may lead to different compression methods. Of course, we have not yet shown how to measure the redundancy in different types of source, and how some redundancies may be more difficult than others to lead to a good compression algorithm. We shall see these from later chapters.

2.7 Compression algorithms

Compression algorithms, in general, aim to convert some source data at the compression end into a compressed message, and to convert it back from the compressed message at the decompression end. This requires making certain assumptions about the source before the conversion takes place. For example, we may need to assume that the source contains symbols from an alphabet and the symbols occur following a specific probability distribution before any coding happens.

Any compression methods involve essentially two types of work: modelling and coding. The model represents our knowledge about the source domain and the manipulation of the source redundancy. The device that is used to fulfil the task of coding is usually called *coder* meaning *encoder*. Based on the model and some calculations, the coder is used to derive a code and encode (compress) the input. A coder can be independent of the model.

A similar structure applies to decoding algorithms. There is again a *model* and a *decoder* for any decoding algorithm.

Such a model-coder structure at both the compression and decompression ends can be seen clearly in Figure 2.16. Here we use *Model-C* to represent the model for the coder and *Model-D* for the decoder. We use symbols `ABC···` to represent the source data and `010001100···` to represent the compressed message.

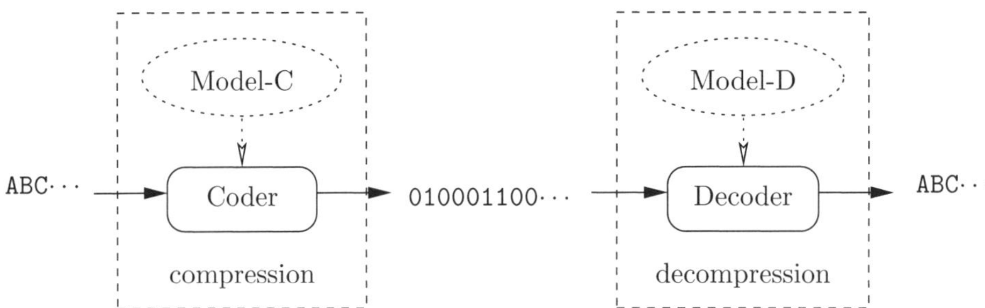

Figure 2.16: Static compression system

Conceptually, we can distinguish *two* types of compression algorithms, namely, *static*, or *adaptive* compression, based on whether the models may be updated during the process of compression or decompression.

- **Static (non-adaptive) system** (Figure 2.16): The model (Model-C or Model-D) remains unchanged during the compression or decompression process.
- **Adaptive system** (Figure 2.17): The model may be changed during the compression or decompression process according to the change of input (or feedback from the output).

  Some adaptive algorithms actually build the model based on the input starting from an empty model.

In practice, a compression software or hardware system often contains a number of static and adaptive algorithms.

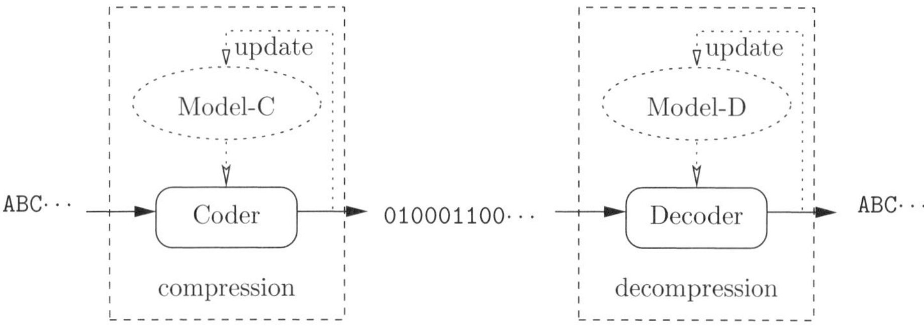

Figure 2.17: Adaptive compression system

In some compression systems, the model for compression (Model-C in the figures) and that for decompression (Model-D) are identical. If they are identical, the compression system is called *symmetric*, otherwise, it is said to be *non-symmetric*. The compression using a symmetric system is called *symmetric compression*, and the compression using an asymmetric system is called *asymmetric compression*.

In terms of the length of codewords used *before* or *after* compression, compression algorithms can be classified into the following categories:

1. **Fixed-to-fixed**: each symbol before compression is represented by a fixed number of bits (for example, 8 bits in ASCII format) and is encoded as a sequence of bits of a fixed length after compression.

   Example 2.33 *A:00, B:01, C:10, D:11* [2]

2. **Fixed-to-variable**: each symbol before compression is represented by a fixed number of bits and is encoded as a sequence of bits of different length.

   Example 2.34 *A:0; B:10; C:101; D:0101.*

3. **Variable-to-fixed**: a sequence of symbols represented in a different number of bits before compression is encoded as a fixed-length sequence of bits.

   Example 2.35 *ABCD:00; ABCDE:01; BC:11.*

4. **Variable-to-variable**: a sequence of symbols represented in a different number of bits before compression is encoded as a variable-length sequence of bits.

   Example 2.36 *ABCD:0; ABCDE:01; BC:1; BBB:0001.*

We will see various types of compression algorithms in the later chapters and a summary of the different types of coding methods in Section 7.6.

[2]For ease of reading, the symbols themselves are used instead of their ASCII codewords.

Summary

Information theory provides a good foundation for compression algorithms. By working on symbolic data, statistical models can be adopted for solving compression problems. Prefix codes, Kraft inequality, and entropy are useful tools in searching efficient codes and identifying the quality of compression algorithms. Symmetric and asymmetric compression models offer different performance and are useful for different types of source. A good compression algorithm often relies upon effective identification of redundancies in the interested source.

Learning outcomes

On completion of your studies in this chapter, you should be able to:

- explain why modelling and coding are usually considered separately for compression algorithm design
- identify the model and the coder in a compression algorithm
- identify prefix codes
- demonstrate the relationship between prefix codes, Kraft inequality and unique decodability
- explain how entropy can be used to measure the code optimum.

Exercises

E2.1 What is a prefix code? What can we conclude about the lengths of a prefix code? Provide an example to support your argument.

E2.2 If a code is *not* a prefix code, can we conclude that it will not be uniquely decodable? Give reasons.

E2.3 Determine whether the following codes are uniquely decodable:

(a) (0,01,11,111)
(b) (0,01,110,111)
(c) (0,10,110,111)
(d) (1,10,110,111).

E2.4 Decide the code efficiency.

Suppose that a source alphabet $\mathcal{S} = (A, B, C, D)$ with probability distribution $\mathcal{P} = (\frac{1}{2}, \frac{1}{4}, \frac{1}{8}, \frac{1}{8})$ is known and a prefix code $\mathcal{C} = (0, 10, 110, 1110)$ is recommended.

Laboratory

L2.1 Design and implement a program method `entropy` which takes a probability distribution (i.e. a set of probability values whose sum is 1) as the argument and returns the first-order entropy of the source.

L2.2 Design and implement a program method `averageLength` which takes two arguments: (1) a set of lengths of a code; (2) the probability distribution of the codewords. It then returns the average length of the code.

Assessment

S2.1 Describe briefly how each of the two classes of lossless compression algorithms, namely the *adaptive* and the *non-adaptive*, works in its model. Illustrate each with an appropriate example.

S2.2 Determine whether the following codes for (A, B, C, D) are *uniquely decodable*. Give your reasons for each case.

(a) (0, 10, 101, 0101)

(b) (000, 001, 010, 011)

(c) (00, 010, 011, 1)

(d) (0, 001, 10, 010).

S2.3 Determine whether the code (0, 10, 011, 110, 1111) is a prefix code and explain why.

S2.4 If a code is a prefix code, what can we conclude about the lengths of the codewords?

Bibliography

[Bri63] L. Brillouin. *Science and Information Theory.* Academic Press, New York, 1963.

[CT05] T.M. Cover and J.A. Thomas. *Elements of Information Theory.* Wiley-Interscience, 2005.

[Gra90] R.M. Gray. *Entropy and Information Theory.* Springer-Verlag, New York, November 1990.

[Ham86] R.W. Hamming. *Coding and Information Theory.* Prentice Hall, Englewood Cliffs, New Jersey, second edition, 1986.

[Jon79] D.S. Jones. *Elementary Information Theory.* Clarendon Press, Oxford, 1979.

[McE77] R. McEliece. *The Theory of Information and Coding.* Addison-Wesley, Reading, MA, 1977.

[Sha48] C.E. Shannon. A mathematical theory of communication. *Bell System Technical Journal*, 27:379–423, 623–656, 1948.

[Sha51] C.E. Shannon. Prediction and entropy of printed English. *Bell System Technical Journal*, 30(1):50–64, January 1951.

[SW49] C.E. Shannon and W. Weaver. *The Mathematical Theory of Communication.* The University of Illinois Press, Urbana, Illinois, 1949.

Chapter 3

Run-length algorithms

In this chapter, we consider a type of redundancy, as in Example 2.24, where a consecutive sequence of symbols can be identified, and introduce a class of simple but useful lossless compression algorithms called run-length algorithms or run-length encoding (RLE for short).

We first introduce the ideas and approaches of the run-length compression techniques. We then move on to show how the algorithm design techniques learnt in Chapter 1 can be applied to solve the compression problem.

3.1 Run-length

The consecutive recurrent symbols are usually called *runs* in a sequence of symbols. Hence the source data of interest is a sequence of symbols from an *alphabet*. The goal of the run-length algorithm is to identify the runs and record the *length* of each run and the symbol in the run.

Example 3.1 *Consider the following strings:*

1. `KKKKKKKKK`
2. `ABCDEFG`
3. `ABABBBC`
4. `abc123bbbbCDE`.

We highlight the runs in each instance by a small shade.

1. `KKKKKKKKK`: There is a run of length 9 on symbol `K`.
2. `ABCDEFG`: There is no run.
3. `ABABBBC`: There is a run of length 3 on symbol `B`.
4. `abc123bbbbCDE`: There is a run of length 4 on symbol `b`.

A run-length algorithm assigns codewords to runs instead of coding individual symbols. The runs are replaced by a tuple (r, l, s) for (run-flag, run-length, run-symbol) respectively, where s is a member of the alphabet of the symbols and r and l are not.

Example 3.2 *String* `KKKKKKKKK`*, containing a run containing 9* `K`*s, can be replaced by triple ('r', 9, 'K'), or a short unit* `r9K` *consisting of the symbol* `r`*,* `9` *and* `K`*, where* `r` *represents the case of 'repeating symbol',* `9` *means '9 times of occurrence' and* `K` *indicates that this should be interpreted as 'symbol* `K`*' (repeating 9 times).*

When there is no run, in `ABCDEFG` for example, the run-flag n is assigned to represent the non-repeating symbols and l, the length of the longest non-recurrent symbols are counted. Finally, the entire non-recurrent string is copied as the third element in the triple. This means that non-repeating string `ABCDEFG` is replaced by ('n', 7, '`ABCDEFG`'), and `n7ABCDEFG` for short.

Run-length algorithms are very effective if the source contains many runs of consecutive symbols. In fact, the symbols can be characters in a text file, 0s and 1s in a binary file, or any composite units such as colour pixels in an image, or even component blocks of larger sound files.

Although simple, run-length algorithms have been used well in practice. The so-called HDC (hardware data compression) algorithm, used by tape drives connected to IBM computer systems, and a similar algorithm used in the IBM System Network Architecture (SNA) standard for data communications are still in use today.

We briefly introduce the HDC algorithm below.

3.2 Hardware data compression (HDC)

For convenience, we will look at a simplified version of the HDC algorithm. In this form of run-length coding, we assume each run or the non-repeating symbol sequence contains no more than 64 symbols. There are two types of control characters. One is a flag for runs and the other is for non-run sequences.

We define the repeating control characters as $r_3, r_4, \cdots, r_{63}$. The subscripts are numbers to indicate the length of the run. For example, r_5 indicates the case of a run of length 5. The coder replaces each sequence of consecutive identical symbols with one of the repeating control characters $r_3, \cdots, r_{63}$ and depends on the run-length followed by the repeating symbol. For example, `VVVV` can be replaced by r_4`V`. For a run of spaces, the algorithm will use the control characters $r_2, r_3, \cdots, r_{63}$ only, but leave out the symbol part. For example, r_7r_4`V` can be decoded as

␣␣␣␣␣␣␣VVVV

For the non-run parts, non-repeating control characters $n_1, n_2, \cdots, n_{63}$ are used which are followed by the length of the longest non-repeating characters until the next run or the end of the entire file. For example, ABCDEFG will be replaced by n_7ABCDEFG.

This simple version of the HDC algorithm essentially uses only ASCII codes for the single symbols, or a total of 123 control characters including a run-length count. Each r_i, where $i = 2, \cdots, 63$, is followed by either another control character or a symbol. If the following symbol is another control character, r_i (alone) signifies i repeating space characters (i.e. spaces or blanks). Otherwise, r_i signifies that the symbol immediately after it repeats i times. Each n_i, where $i = 1, \cdots, 63$, is followed by a sequence of i *non-repeating* symbols.

Applying the following 'rules', it is easy to understand the outline of the *encoding* and *decoding* run-length algorithms below.

3.2.1 Encoding

Repeat the following until the end of input file:

Read the source (e.g. the input text) symbols sequentially and

1. if a string[1] of i ($i = 2, \cdots, 63$) consecutive spaces is found, output a single control character r_i
2. if a string of i ($i = 3, \cdots, 63$) consecutive symbols other than spaces is found, output two characters: r_i followed by the repeating symbol
3. otherwise, identify a longest string of $i = 1, \cdots, 63$ non-repeating symbols, where there is no consecutive sequence of two spaces or of three other characters, and output the non-repeating control character n_i followed by the string.

Example 3.3 GGG␣␣␣␣␣␣BCDEFG␣␣55GHJK␣LM777777777777
can be compressed to r_3Gr_6n_6BCDEFG$r_2n_9$55GHJK␣LMr_{12}7.

Solution

1. The first three Gs are read and encoded by r_3G.
2. The next six spaces are found and encoded by r_6.
3. The non-repeating symbols BCDEFG are found and encoded by n_6BCDEFG.
4. The next two spaces are found and encoded by r_2.
5. The next nine non-repeating symbols are found and encoded by $n_9$55GHJK␣LM.
6. The next twelve '7's are found and encoded by r_{12}7.

Therefore the encoded output is: r_3Gr_6n_6BCDEFG$r_2n_9$55GHJK␣LMr_{12}7.

[1] i.e. a sequence of symbols.

3.2.2 Decoding

The decoding process is similar to that for encoding and can be outlined as follows:

> Repeat the following until the end of input coded file:
>
> Read the codeword sequence sequentially and
>
> 1. if an r_i is found, then check the next codeword
>   (a) if the codeword is a control character output i spaces
>   (b) otherwise output i (ASCII codes of) repeating symbols
> 2. otherwise, output the next i non-repeating symbols.

Observation

It is not difficult to observe from a few examples that the performance of the HDC algorithm (as far as the compression ratio concerns) is:

- excellent[2] when the source contains many runs of consecutive symbols
- poor when there are many segments of non-repeating symbols.

Therefore, run-length algorithms are often used as a subroutine in other more sophisticated coding.

3.3 Algorithm Design

We have so far learnt the ideas behind the HDC algorithm as well as run-length algorithms in general. To learn how to design our own compression algorithms, we look at how to derive a simple version of HDC applying the algorithm design techniques introduced in Chapter 1.

Stage 1: Description of the problem

A problem is a general question to be answered. However, a question may be too general to lead to an algorithmic solution or too vague to even understand the issues involved. To help us understand the HDC problem better, we look at Example 3.1 again.

From the example, we study the input-output to reflect the behaviour of the algorithm to be developed. It becomes clear to us soon that a run can be described by two parts as a pair (c, s), where c represents the control character with a count, and s the repeating symbol or non-run string depending on whether c is r_i or n_i.

[2]It can be even better than entropy coding such as Huffman coding.

Compression

The output of each instance in the example can then be written:

1. input: KKKKKKKKK, output: r_9K, or (r_9, K)
2. input: ABCDEFG, output: n_7ABCDEFG
3. input: ABABBBC, output: n_3ABAr_3Bn_1C
4. input: abc123bbbbCDE, output: $n_6$abc123r_4bn_3CDE.

Decompression

1. input: r_9K output: KKKKKKKKK
2. input: n_7ABCDEFG, output: ABCDEFG
3. input: n_3ABAr_3Bn_1C, output: ABABBBC
4. input: $n_6$abc123r_4bn_3CDE. output: abc123bbbbCDE.

Since the compression process is reversible, the RLE scheme is lossless compression.

We now write the description of the problem:

Problem: Find and replace each run by a hopefully shorter codeword to indicate run-flag, run-length and run-symbol respectively.
Input: A sequence of symbols from an alphabet.
Output: Another hopefully shorter sequence of symbols.

Stage 2: Mathematical modelling

The mathematical model for run-length is the so-called *Markov model.* Named after the Russian mathematician Andrei Andrevich Markov (1856–1922), Markov models are popular in describing data with certain dependence. Models are useful for estimating the entropy and for comparison with other methods. However, we shall focus on the coder design in this chapter instead of discussing the details about the model. We will look more closely at the data and decide how the consecutive recurrent symbols can be represented in a shorter string.

Let us analyse four instances in Example 3.1 first. The symbols in each sequence are read one by one in order to identify runs and their lengths. For convenience, the current symbol is highlighted, and the control codewords r_i or n_i are used to mean a run or non-run of length i respectively. We will play with a few examples to gain better understanding of the problem.

Compression

1. In this instance, the whole input string is a run, which can be identified by reading the characters one at a time, and comparing it with the previous symbol.

String	Repeat	Count	Non-repeat
KKKKKKKKK	K	0	
KKKKKKKKK	KK	1	
KKKKKKKKK	KK	2	
KKKKKKKKK	KK	3	
KKKKKKKKK	KK	4	
KKKKKKKKK	KK	5	
KKKKKKKKK	KK	6	
KKKKKKKKK	KK	7	
KKKKKKKKK	KK	8	
result	r9K	9	

2. This is an instance where there is no run in the data.

String	Repeat	Non-repeat	Count
ABCDEFG	A	0	
ABCDEFG		AB	1
ABCDEFG		ABC	2
ABCDEFG		ABCD	3
ABCDEFG		ABCDE	4
ABCDEFG		ABCDEF	5
ABCDEFG		ABCDEFG	6
result		ABCDEFG	7

3. This is an instance consisting of one run.

String	Repeat	Non-repeat	Repeat	Non-repeat
ABABBBC	A			
ABABBBC		AB		
ABABBBC		ABA		
ABABBBC		ABAB		
ABABBBC		ABA	1BB	
ABABBBC		ABA	2BB	
ABABBBC		ABA	3BC	
result		ABA	r3B	C

4. This is again an instance consisting of one run. However, the process of identifying the run seems more complex than the previous instance.

String	Repeat	Non-repeat	Repeat	Non-repeat
abc123bbbbCDE	a			
abc123bbbbCDE		ab		
abc123bbbbCDE		abc		
abc123bbbbCDE		abc1		
abc123bbbbCDE		abc12		
abc123bbbbCDE		abc123		
abc123bbbbCDE		abc123b		
abc123bbbbCDE		abc123	**1**bb	
abc123bbbbCDE		abc123	**2**bb	
abc123bbbbCDE		abc123	**3**bC	
abc123bbbbCDE		abc123	**4**bC	
abc123bbbbCDE		abc123	**4**b	CD
abc123bbbbCDE		abc123	**4**b	CDE
result		abc123	**r4**b	CDE

The complication in this instance is due to a slightly larger symbol alphabet and a mixture of both cases of characters and digits (a, b, c, C, D, E, 1, 2, 3). Note the count numbers are in bold in order to distinguish themselves from the symbol 1, 2, 3.

Decompression

This is the reverse process of compression. We therefore take the output of the compression as the input string.

1. input: r9K

String	Repeat	Non-repeat
r9K		
r9K		
r9K	KKKKKKKKK	
output	KKKKKKKKK	

2. input: ABCDEFG

String	Repeat	Non-repeat
ABCDEFG		A
ABCDEFG		B
ABCDEFG		C
ABCDEFG		D
ABCDEFG		E
ABCDEFG		F
ABCDEFG		G
output		ABCDEFG

3. input: `ABAr3BC`

String	Repeat	Non-repeat	Repeat	Non-repeat
`ABAr3BC`		`A`		
`ABAr3BC`		`B`		
`ABAr3BC`		`A`		
`ABAr3BC`				
`ABAr3BC`				
`ABAr3BC`			`BBB`	
`ABAr3BC`				`C`
output		`ABABBBC`		

4. input: `abc123r4bCDE`

String	Repeat	Non-repeat	Repeat	Non-repeat
`abc123r4bCDE`		`a`		
`abc123r4bCDE`		`b`		
`abc123r4bCDE`		`c`		
`abc123r4bCDE`		`1`		
`abc123r4bCDE`		`2`		
`abc123r4bCDE`		`3`		
`abc123r4bCDE`				
`abc123r4bCDE`				
`abc123r4bCDE`			`bbbb`	
`abc123r4bCDE`				`C`
`abc123r4bCDE`				`D`
`abc123r4bCDE`				`E`
output		`abc123bbbbCDE`		

We now summarise our model which consists of

1. A finite alphabet of symbols $\mathcal{S} = (s_1, s_2, \cdots, s_n)$, where n is the size of the alphabet.
2. Two states: *repeating* and *non-repeating*.
3. A control character c to indicate a run r_i or not n_i with the counter i to compute the length of the runs or non-runs.
4. A run-length code $\mathcal{C} = (c_1, c_2, \cdots, c_n)$, where each codeword contains a control character followed by a symbol or a string.

Stage 3: Design of the algorithm

We now consider the design of the HDC algorithm.

The best way to develop an algorithm is to take a top-down approach. We usually write down the main idea of the algorithm before adding details. The advantage of this approach is to avoid wasting time on coding for a flawed algorithm.

Compression

Suppose the string is read one symbol at a time. Let *sp*, *sq* and *sr* be three consecutive symbols that are read most recently. The values of three consecutive symbols can be used to detect runs and non-runs.

If the string of symbols is stored in an array S, then each symbol can be accessed via an index. For example, reading *sp* from the array means an assignment $sp \leftarrow S[p]$ and results in $sp = S[p]$, and similarly, reading the next symbol sq is equivalent to $sq \leftarrow S[q]$, where $q = p+1$ and $r = q+1 = p+2$ (Figure 3.1).

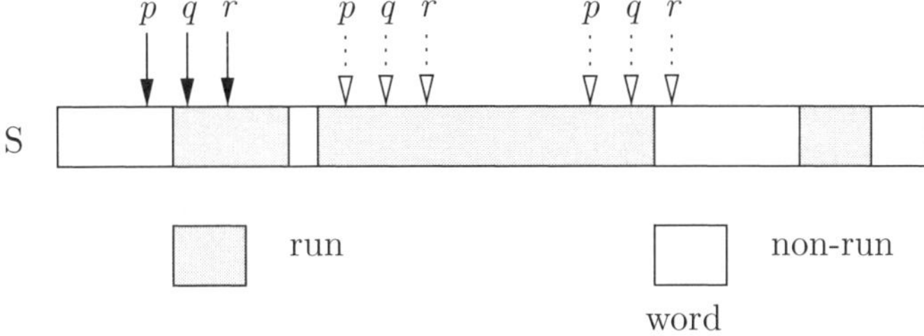

Figure 3.1: Three consecutive variables $S[p]$, $S[q]$ and $S[r]$

One way to concentrate on the top-level structure is to define procedures or functions that can be extended later on. We use $xxx()$ to represent a procedure or function of name xxx.

All the possibilities of runs and non-runs in three consecutive postions can be classified as four cases: they are XXX, XXY, XYY and XYZ, where X, Y and Z represent three different symbols. We look closely at each case and summarise the actions required for each case below:

case	space	actions
XXX		begin_run(), find_longest_run()
XXY	␣␣Y	output_run(r_2)
	else	begin_nonrun(), find_longest_nonrun()
XYY	X␣␣	output_nonrun(n_1, X)
	else	begin_nonrun(), find_longest_nonrun()
XYZ		begin_nonrun(), find_longest_nonrun()

We now define the procedures or functions required:

procedure or function	description
(boolean) begin_run()	true if a run begins
longest_run()	to find the run length
(boolean) begin_nonrun()	not begin_run()
longest_nonrun()	to find the length of a non-run
output_run()	output a run
output_nonrun()	output a non-run
shift1()	$sp \leftarrow sq$, $sq \leftarrow sr$, read sr
shift2()	$sp \leftarrow sr$, read sq, read sr

We also define a string *word* to accumulate a non-run, and a counter *length* to record the current length of a repeating or non repeating string.

We can now write the main algorithm as follows. Note: here the addition sign '+' means concatenation for string operants.

Algorithm 3.1 myHDC encoding

```
INPUT:    symbol string in an array
OUTPUT:   codeword sequence
read sp, sq, sr {i.e. S[p], S[q], S[r] in Figure 3.1}
while not EOF do
   if begin_run() then
      find_longest_run()
      output_run(length, symbol)
   else
      find_longest_nonrun()
      output_nonrun(length, word)
   end if
end while
```

We now work on function *begin_run*().

A run has to satisfy the following conditions:

1. two or more consecutive spaces, or
2. three or more consecutive other symbols in the alphabet.

Algorithm 3.2 (Function) begin_run()

```
INPUT:
RETURN:   true or false
return (sp = sq = '␣') or (sp = sq = sr )
```

Similarly, a non-run needs to contain

1. one space, or
2. two consecutive identical symbols other than spaces, or
3. two different consecutive symbols.

However, since a string is a non-run if it is not a run, this function is actually not needed explicitly for our algorithm.

The other procedures and functions are as follows:

Algorithm 3.3 (Procedure) shift1()

```
INPUT:
OUTPUT:
sp ← sq
sq ← sr
read sr
```

Algorithm 3.4 (Procedure) shift2()

```
INPUT:
OUTPUT:
sp ← sr
read sq
read sr
```

We now move on to the two functions of computing the run length and non-run length. For non-run length, we also need to accumulate the non-run sequence in *word*.

Algorithm 3.5 (Procedure) find_longest_run()

```
INPUT:
OUTPUT:   runLength, symbol
length ← 2 {two spaces only}
if sp = sq = '␣' and sq ≠ sr then
   symbol = '␣'
else {three or more repeating symbols}
   while not EOF and sp = sq = sr do
      runLength ← runLength + 1
      read sr
   end while
   symbol ← sp
end if
if not EOF then
   shift2()
end if
```

Algorithm 3.6 (Procedure) find_longest_nonrun()

```
INPUT:
OUTPUT:   nonrunLength, symbol
word ← sp + sq
length ← 2
while not EOF and [(sp ≠ sq) or (sp = sq ≠ '␣' and sq ≠ sr)] do
    length ← length + 1
    shift1()
    word ← word + sq
end while
if not EOF then
    length ← length − 2
    word ← word − sp − sq
end if
```

Algorithm 3.7 (Procedure) output_run(length, symbol)

```
INPUT:    runlength, symbol
OUTPUT:   codeword for a run
if symbol = '␣' then
    output r_length
else
    output r_length + symbol
end if
```

Algorithm 3.8 (Procedure) output_nonrun(length, word)

```
INPUT:    runlength, word
OUTPUT:   codeword for a non-run
output n_length + word
```

Observation

We have gained some useful insight of algorithm design from the above process.

1. Identify smallest set of symbols to begin with;
2. It is often easier to take a standard *top-down design* approach;
3. It is useful to draw a diagram to help understand the problem.

Decompression

Similarly, we collect the ideas and write them in pseudocode.

myHDC decompression idea:

Algorithm 3.9 myHDC decoding

```
INPUT:    run-length codeword sequence
OUTPUT:   symbol sequence
controlSymbol ← read_next_symbol()
while not EOF do
   if controlSymbol = 'r'_k then
      nextSymbol ← read_next_symbol()
      if nextSymbol is a control symbol then
         output(k, '␣')
      else
         output_run(k, nextSymbol)
      end if
   else
      output_nonrun(k)
   end if
   if nextSymbol is not a control symbol then
      nextSymbol ← read_next_symbol()
   end if
   controlSymbol ← nextSymbol
end while
```

Algorithm 3.10 (Procedure) output_run()

```
INPUT:    k, symbol
OUTPUT:   symbol sequence
for i = 1, i ≤ k, i = i + 1 do
   output symbol
end for
```

Algorithm 3.11 (Procedure) output_nonrun()

```
INPUT:    k
OUTPUT:   symbol sequence
for i = 1, i ≤ k, i = i + 1 do
   symbol ← read_next_symbol()
   output symbol
end for
```

Algorithm design is a complex process. The first version of any algorithm almost certainly contains flaws. As a student, you should not be disappointed or surprised to find errors in your (or anyone else's) algorithms. It is more important to know what causes the problem and how to debug it.

Stage 4: Verification of the algorithm

Correctness

The first thing to verify at this stage is the correctness of the algorithm. We need to make sure that the algorithm produces the expected solutions in all cases. Unfortunately, it is not always easy to give mathematical proof in algorithm analysis. Existing testing methods can, strictly speaking, show only the presence of bugs, but not their absence. Nevertheless, testing is essential in maintaining the quality of software. It is an effective way to find the bugs in an algorithm.

In this book, we take a simple approach that is similar to the so-called *black-box* method. The black-box approach means testing an algorithm or program based only on its interface, without knowing its internal structure. We first prepare a set of test data as inputs to the algorithm and work out the expected outputs from the algorithm for each instance of the inputs. Next we run the algorithm on the test data to get the actual output. Finally we compare the actual output to the expected output.

It is important to design a good set of testing data. The general requirement to test data includes: being easy to check for correctness, being as representative as possible of the real input data to the algorithm, covering extreme and possible illegal values of the input data.

For example, string n_3`GGR`r_3`K`r_2n_5`GHEEN` is not bad for a quick check on Algorithm 3.1 and 3.9 because it covers most working cases in the algorithms.

Saving percentage

Next, we want to estimate the compression quality at this stage.

Let the source file contain N symbols from an alphabet $s_1, s_2, \cdots, s_n$. Suppose there are possible runs of length $l_1, l_2, \cdots, l_m$. The number of runs of length l_i is k_i, where $i = 1, \cdots, m$, where $m \leq N$. (In practice, m would be much smaller than N.)

Hence the average length of the runs is

$$\bar{l} = \frac{\sum_{i=1}^{m} k_i l_i}{\sum_{i=1}^{m} k_i}$$

where the total number of runs is $M = \sum_{i=1}^{m} k_i$.

If we know the probability distribution of the run lengths in a source $(p_1, p_2, \cdots, p_m)$, we can estimate $\bar{l}$ by

$$\bar{l} = \sum_{i=1}^{m} p_i l_i$$

Since each codeword is a tuple (r, l, s), the codewords are of fixed length. For example, if $0 \leq l \leq 255$, then the length for each codeword is 3 bytes. That is to say, each codeword is equivalent to three symbols long.

Therefore, the saving percentage is

$$1 - \frac{M(\bar{l} - 3)}{N}$$

As we can see, the performance of the run-length algorithm depends on the number and average length of the runs. Run-length algorithms are very effective if the data source contains many runs of consecutive symbol. The symbols can be characters in a text file, 0s or 1s in a binary file or black-and-white pixels in an image.

One easy way to verify an algorithm is to feed the algorithm data and verify expected outcomes. For example, the previous example of data can be the input to the algorithms. We leave this as an exercise to the reader.

Stage 5: Estimation of the computational complexity

Time efficiency

We next move on to the time complexity of the algorithms.

From the encoding algorithms, we see it is of $O(N)$ where N is the number of symbols in the source file.

In theory, one would justify the algorithmic solutions before moving on to the implementation stage. Unfortunately, in modern practice, it is easy for people to start implementation early despite obvious disadvantages of the approach. We hope that we have in this section demonstrated a systematic approach to algorithm design. The reader is encouraged to apply this approach to all the exercises in the book.

Summary

Run-length algorithms are simple, fast and effective for a source that contains many long runs. The HDC algorithm is a good example of such an approach which is still used today. One good way of studying algorithms is to play with examples.

Learning outcomes

On completion of your studies in this chapter, you should be able to:

- describe a simple version of run-length algorithm
- explain how a run-length algorithm works
- explain under what conditions a run-length algorithm may work effectively
- explain, with an example, how the HDC algorithm works
- derive a new version of a run-length algorithm following the main steps of algorithm design process.

Exercises

E3.1 Apply the HDC (hardware data compression) algorithm to the following sequence of symbols:

`kkkk␣␣␣␣␣␣␣␣␣␣␣␣g␣␣hh5522777666abbbbcmmj␣␣##`

Show the compressed output and explain the meaning of each control symbol.

E3.2 Explain how the compressed output from the above question can be reconstructed using the decompression algorithm.

E3.3 Provide an example of a source file on which the HDC algorithm would perform very badly.

E3.4 Outline the main stages of an algorithm design, using a simplified version of the run-length algorithm.

Laboratory

L3.1 Based on the outline of the simple HDC algorithm, derive your version of the HDC algorithm in pseudocode which allows an easy implementation in your favourite program language.

L3.2 Implement your version of the HDC algorithm. Use `MyHDC` as the name of your main class/program.

L3.3 Provide two source files `good.source` and `bad.source`, on which HDC would perform very well and very badly respectively. Indicate your definition of 'good' and 'bad' performance.

Hint: Define the input and output of your (compression and decompression) algorithms first.

Assessment

S3.1 Describe with an example how a run-length coder works.

S3.2 Apply the HDC (hardware data compression) algorithm to the sequence:

`␣␣␣␣␣␣␣BC␣␣␣A␣1144330000␣␣EFGHHHH`

Demonstrate the compressed output and explain the meaning of each control symbol.

Bibliography

[Cap59] J. Capon. A Probabilistic Model for Run-Length Coding of Pictures. *IRE Transactions on Information Theory*, pages 157–163, 1959.

[Gol66] S.W. Golomb. Run-length Encodings. *IEEE Transactions on Information Theory*, IT-12(3):399–401, 1966.

[Sal04] D. Salomon. *Data Compression: the Complete Reference.* Springer, 2004.

[Say00] K. Sayood. *Introduction to Data Compression.* Morgan Kaufmann, 2000.

Chapter 4

Huffman coding

In this chapter, we formally study the Huffman coding algorithms and apply the theory in Chapter 2.

4.1 Static Huffman coding

Huffman coding is a successful compression method used originally for text compression. In any text, some characters occur far more frequently than others. For example, in English text, the letters E, A, O, T are normally used much more frequently than J, Q, X.

Huffman's idea is, instead of using a fixed-length code such as 8 bit extended ASCII or DBCDIC for each symbol, to represent a frequently occurring character in a source with a shorter codeword and to represent a less frequently occurring one with a longer codeword. Hence the total number of bits of this representation is significantly reduced for a source of symbols with different frequencies. The number of bits required is reduced for each symbol *on average*.

Compression

In order to understand the problem, we first look at some examples of source texts.

Example 4.1 *Consider the string BILL BEATS BEN. For convenience, we ignore the two spaces.*

The frequency of each symbol is:

```
B I L E A T S N
3 1 2 2 1 1 1 1
```

Sort the list by frequency:

```
B L E I A T S N
3 2 2 1 1 1 1 1
```

This source consists of symbols from an alphabet (B, L, E, I, A, T, S, N) with the recurrent statistics (3, 2, 2, 1, 1, 1, 1, 1). We want to assign a variable length of prefix code to the alphabet, i.e. one codeword of some length for each symbol.

The input of the compression algorithm is a string of text. The output of the algorithm is the string of binary bits to interpret the input string. The problem contains three subproblems:

1. Read input string
2. Interpret each input symbol
3. Output the codeword for each input symbol.

The first and the last subproblems are easy. For the first subproblem we only need a data structure to allow the access to each symbol one after another. Suppose the prefix code is $\mathcal{C} = (c_1, c_2, \cdots, c_n)$, where $n = 8$ in this example. For the last subproblem, we only need a means to find the corresponding codeword for each symbol and output the codeword.

So we focus on the second subproblem which is how to derive the code $\mathcal{C}$.

We write the description of the problem:

Main Subproblem: Derive an optimal or suboptimal prefix code.
Input: An alphabet and a frequency table.
Output: A prefix code $\mathcal{C}$ such that the average length of the codewords is as short as possible.

Modelling is fairly easy if we follow the statistical model in Chapter 2. The alphabet of a source is $\mathcal{S} = (s_1, s_2, \cdots, s_n)$ which associates with a probability distribution $\mathcal{P} = (p_1, p_2, \cdots, p_n)$. Note a frequency table can be easily converted to a probability table. For example, we use the previous example where a frequency table (3, 2, 2, 1, 1, 1, 1, 1) is given for alphabet (B, L, E, I, A, T, S, N). The total frequency is $3 + 2 + 2 + 1 + 1 + 1 + 1 + 1 = 12$. The probability for each symbol is the ratio of its frequency over the total frequency. We then have the probability distribution for prediction of a source in the future $(\frac{3}{12}, \frac{2}{12}, \frac{2}{12}, \frac{1}{12}, \frac{1}{12}, \frac{1}{12}, \frac{1}{12}, \frac{1}{12})$.

We now consider how to construct a prefix code in which short codewords are assigned to frequent symbols and long codewords to rare symbols. Recall in Section 2.3.3 that any prefix code can be represented in a 0-1 tree where all the symbols are at leaves and the codeword for each symbol consists of the collection of the 0s and 1s from the root to that leaf (Figure 2.5). The short codewords are at lower level leaves and the long codewords at higher level leaves (Figure 2.8).[1] If we have such a binary tree for the alphabet, we have the prefix code for the source.

Suppose the prefix code is $\mathcal{C} = (c_1, c_2, \cdots, c_n)$ with lengths $\mathcal{L} = (l_1, l_2, \cdots, l_n)$ respectively.

Our problem of deriving a prefix code becomes a problem of how to construct a 0-1 tree so that

[1] Note the root is at the lowest level of the tree.

1. All the symbols are leaves
2. If $p_j > p_i$, then $l_j \leq l_i$, for all $i, j = 1, \cdots, n$
3. Two longest codewords are identical except for the last bit.

For example, symbol B has a higher frequency than L, therefore the codeword length for B should be no longer than L.

The longest codewords should be assigned to the more rare symbols which are the last two symbols in our sorted list:

```
B L E I A T S N
3 2 2 1 1 1 1 1
```

If the codeword for S is 0000, then the codeword for N should be 0001.

There are two approaches to construct a binary tree: one is starting from the leaves to build the tree from the bottom up to the root. This 'bottom-up' approach is used in Huffman encoding. The other is starting from the root down to the leaves. The 'top-down' approach is used in Shannon-Fano encoding (Section 4.2).

4.1.1 Huffman approach

We first look at Huffman's 'bottom-up' approach. Here we begin with a list of symbols as the tree leaves. The symbols are repeatedly combined with other symbols or subtrees, two items at a time, to form new subtrees. The subtrees grow in size by combination on each iteration until the final combination before reaching the root.

In order to easily find the two items with the smallest frequency, we maintain a *sorted* list of items in descending order. With minor changes, the method also works if an ascending order list is maintained.

Figure 4.1 shows how the tree is built from the leaves to the root step by step. It carries out the following steps in each iteration:

1. Combine the last two items which have the minimum frequencies or probabilities on the list and replace them by a combined item.
2. The combined item, which represents a subtree, is placed accordingly to its combined frequency on the sorted list.

For example, in Figure 4.1(1), the two symbols S and N (in shade) with the least frequencies are combined to form a new combined item SN with a frequency 2. This is the frequency sum of two singleton symbols S and N. The combined item SN is then inserted to the second position in Figure 4.1(2) in order to maintain the sorted order of the list.

Note there may be more than one possible place available. For example, in Figure 4.1(2), SN with a frequency of 2 can also be inserted immediately before symbol I, or before E. In this case, we always place the newly combined item to a *highest* possible position to avoid it getting combined again too soon. So SN is placed before L.

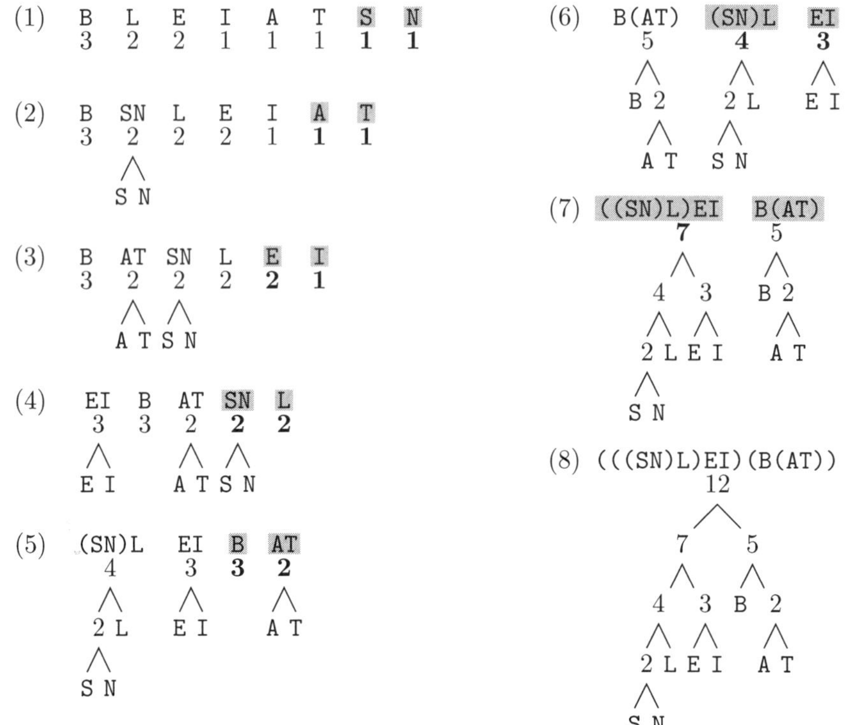

Figure 4.1: Building a Huffman tree

Generalising from the example, we derive the following algorithm for building the tree:

Building the binary tree

Sort alphabet in descending order $\mathcal{S} = (s_1, s_2, \cdots, s_n)$ according to the associated probability distribution. $\mathcal{P} = (p_1, p_2, \cdots, p_n)$. Each s_i represents the root of a subtree. Repeat the following until there is only one composite symbol in S:

1: If there is one symbol, the tree is the root and the leaf. Otherwise, take two symbols s_i and s_j in the alphabet which have the lowest probabilities p_i and p_j.

2: Remove s_i and s_j from the alphabet and add a new combined symbol (s_i, s_j) with probability $p_i + p_j$. The new symbol represents the root of a subtree. Now the alphabet contains one fewer symbol than before.

3: Insert the new symbol (s_i, s_j) to a highest possible position so the alphabet remains the descending order.

Generating the prefix code

Once we have the binary tree, it is easy to assign a 0 to the left branch and a 1 to the right branch for each internal node of the tree as in Figure 4.2. The 0-1 values marked next to the edges are usually called the *weights* of the tree. A tree with these 0-1 labels is called a *weighted* tree. The weighted binary tree derived in this way is called a *Huffman tree.*

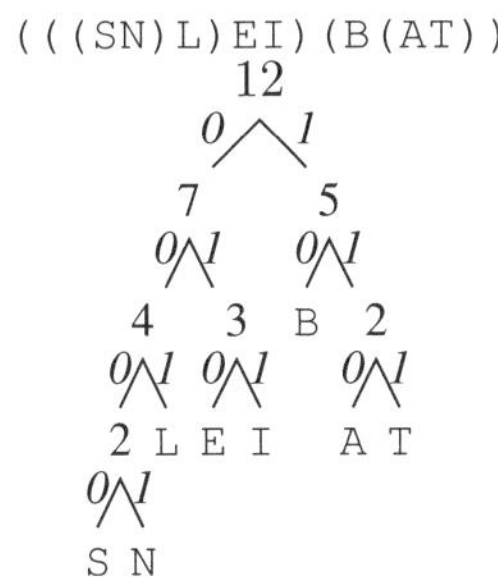

Figure 4.2: A Huffman tree

We then, for each symbol at a leaf, collect the 0 or 1 bit while traversing each tree path from the root to the leaf. When we reach a leaf, the collection of the 0s and 1s forms the prefix code for the symbol at that leaf. The codes derived in this way are called *Huffman codes.*

For example, the collection of the 0s and 1s from the root to leaf for symbol `E` is first a left branch 0, then right branch 1 and finally left branch 0. Therefore the codeword for symbol `E` is 010. Traversing in this way for all the leaves, we derive the prefix code (10 001 010 011 110 111 0000 0001) for the whole alphabet (`B, L, E, I, A, T, S, N`) respectively. A prefix code generated in this way is called *Huffman code.*

4.1.2 Huffman compression algorithm

We first outline the ideas of Huffman compression algorithm with missing details.

Algorithm 4.1 Huffman encoding ideas

1: Build a binary tree where the leaves of the tree are the symbols in the alphabet.
2: The edges of the tree are labelled by a 0 or 1.
3: Derive the Huffman code from the Huffman tree.

This algorithm is easy to understand. In fact, the process of labelling the 0s and 1s does not have to be at the end of construction of the entire Huffman tree. An assignment of a 0 or 1 can be fulfilled as soon as two items are combined, beginning from the least significant bit of each codeword.

We now add details and derive an algorithm as follows.

Algorithm 4.2 Huffman encoding

INPUT: a sorted list of one-node binary trees $(t_1, t_2, \cdots, t_n)$ for alphabet $(s_1, \cdots, s_n)$ with frequencies $(w_1, \cdots, w_n)$

OUTPUT: a Huffman code with n codewords

initialise a list of one-node binary trees $(t_1, t_2, \cdots, t_n)$ with weight $(w_1, w_2, \cdots, w_n)$ respectively
for $k = 1$; $k < n$; $k = k + 1$ **do**
  take two trees t_i and t_j with minimal weights $(w_i \leq w_j)$
  $t \leftarrow merge(t_i, t_j)$ with weight $w \leftarrow w_i + w_j$,
  where $left\_child(t) \leftarrow t_i$ and $right\_child(t) \leftarrow t_j$
  $edge(t, t_i) \leftarrow 0$; $edge(t, t_j) \leftarrow 1$
end for
output every path from the root of t to a leaf, where $path_i$ consists of consecutive edges from the root to $leaf_i$ for s_i

Figure 4.3 shows an example of how this practical approach works step by step.

Canonical and minimum-variance Huffman coding

We have followed the two 'rules' below as standard practice during the derivation of a Huffman tree in this section:

1. A newly created item is placed at the highest possible position in the alphabet list while keeping the list sorted.
2. When combining two items, the one higher up on the list is assigned 0 and the one lower down 1.

The Huffman code derived from a process that follows these rules is called a *canonical and minimum-variance code.* The code is regarded as standard and the length difference among the codewords is kept to the minimum. Huffman coding that follows these rules is called *canonical and minimum-variance Huffman coding.*

Note the canonical and minimum-variance Huffman code is *not* necessarily unique for a given alphabet with associated probability distribution, because there may be more than one way to sort the alphabet list. For example, alphabet (B, L, E, I, A, T, S, N) with the probabilities (3, 2, 2, 1, 1, 1, 1) may be sorted in many ways. Figure 4.4 shows two different canonical and minimum-variance Huffman trees for the same source, one is based on (B, L, E, I, A, T, S, N) and the other on (B, L, E, I, A, S, T, N) with the only difference in the position of symbols `T` and `S` (see highlighted symbols in both lists).

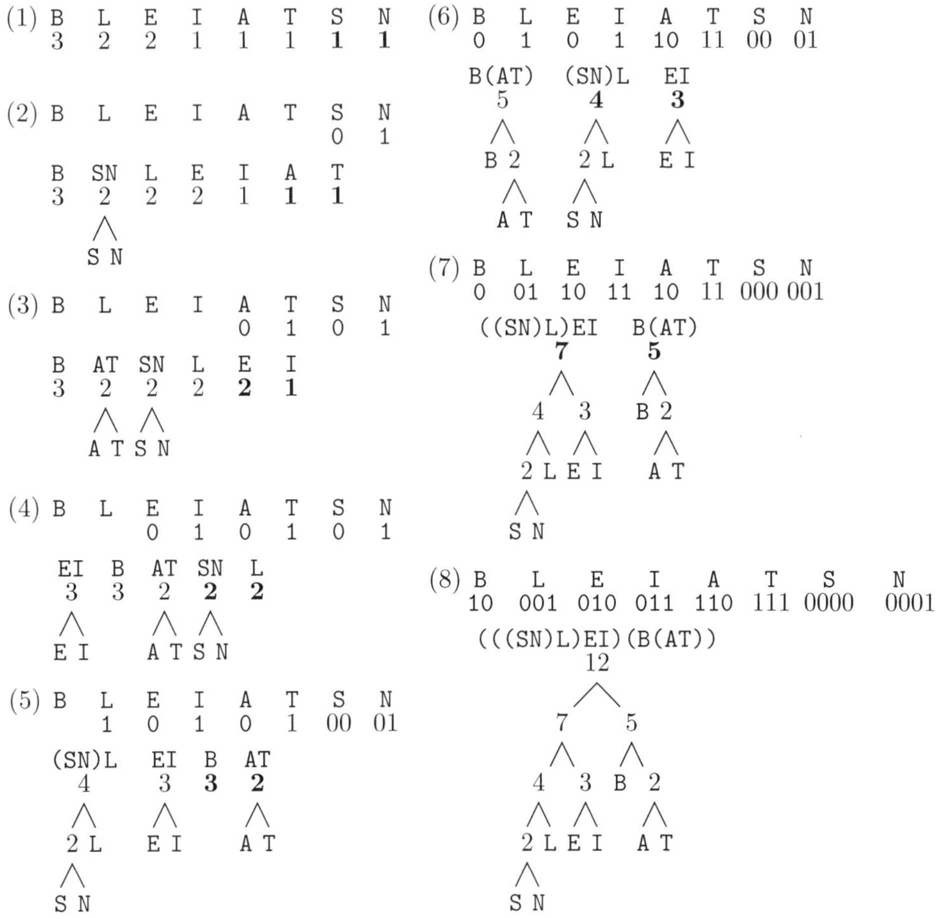

```
(1) B  L  E  I  A  T  S  N
    3  2  2  1  1  1  1  1

(2) B  L  E  I  A  T  S  N
                      0  1

    B  SN  L  E  I  A  T
    3  2   2  2  1  1  1
       /\
       S N

(3) B  L  E  I  A  T  S  N
                0  1  0  1

    B  AT  SN  L  E  I
    3  2   2   2  2  1
       /\  /\
       A T S N

(4) B  L  E  I  A  T  S  N
          0  1  0  1  0  1

    EI  B  AT  SN  L
    3   3  2   2   2
    /\     /\  /\
    E I    A T S N

(5) B  L  E  I  A  T  S   N
       1  0  1  0  1  00  01

    (SN)L  EI  B  AT
      4    3   3  2
      /\   /\     /\
     2 L   E I    A T
     /\
     S N

(6) B  L  E  I  A   T   S   N
    0  1  0  1  10  11  00  01

    B(AT)  (SN)L  EI
      5      4     3
      /\     /\    /\
     B 2    2 L    E I
       /\   /\
       A T  S N

(7) B  L   E   I   A   T   S    N
    0  01  10  11  10  11  000  001

    ((SN)L)EI   B(AT)
        7         5
        /\        /\
       4  3      B 2
      /\  /\       /\
     2 L  E I      A T
     /\
     S N

(8) B   L    E    I    A    T    S     N
    10  001  010  011  110  111  0000  0001

    (((SN)L)EI)(B(AT))
           12
          /  \
         7    5
        /\    /\
       4  3  B  2
      /\  /\    /\
     2 L  E I   A T
     /\
     S N
```

Figure 4.3: Deriving a Huffman code

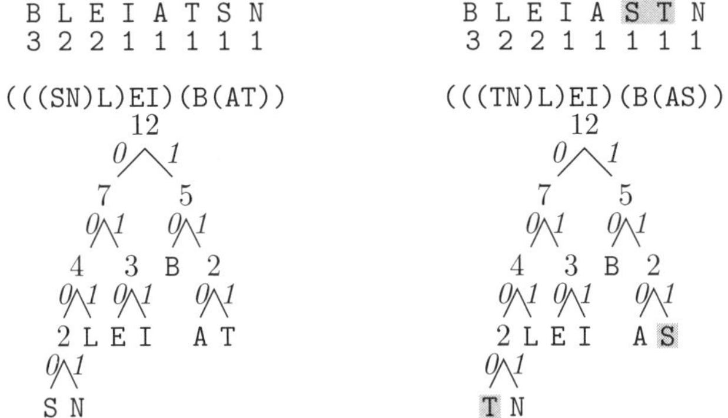

Figure 4.4: Two canonical and minimum-variance trees

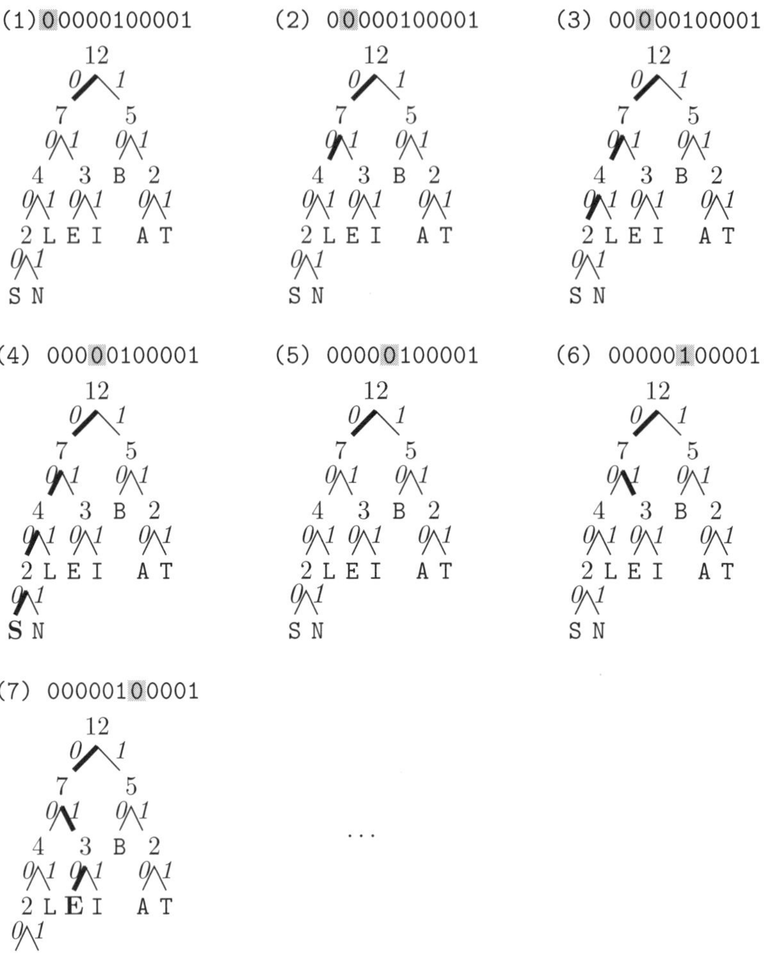

Figure 4.5: Huffman decompression process

4.1.3 Huffman decompression algorithm

The decompression algorithm involves the operations where the codeword for a symbol is obtained by 'walking' down from the root of the Huffman tree to the leaf for each symbol.

Example 4.2 *Decode the sequence 00000100001 using the Huffman tree in Figure 4.2.*

Figure 4.5 shows the first seven steps of decoding the symbols `S` and `E`. The decoder reads the 0s or 1s bit by bit. The 'current' bit is highlighted in shade in the sequence to be decompressed on each step. The edge chosen by the decompression algorithm is marked as a bold line. For example, in step (1), starting from the root of the Huffman tree, we move along the left branch one edge down to the left child since a bit 0 is read. In step (2), we move along the left branch again to the left child since a bit 0 is read, and so on. When we reach a leaf, for example, in step (4), the symbol (the bold 'S') at the leaf is output. This process starts from the root again (5) until step (7) when another leaf is reached and the symbol 'E' is output.

The decoding process ends when `EOF` is reached for the entire string.

We now outline the ideas of Huffman decoding.

Algorithm 4.3 Huffman decoding ideas

1: Read the coded message bit by bit. Starting from the root, we traverse one edge down the tree to a child according to the bit value. If the current bit read is 0 we move to the left child, otherwise, to the right child.
2: Repeat this process until we reach a leaf. If we reach a leaf, we will decode one character and restart the traversal from the root.
3: Repeat this read-and-move procedure until the end of the message.

Adding more details, we have the following algorithm:

Algorithm 4.4 Huffman decoding

INPUT: a Huffman tree and a 0-1 bit string of encoded message
OUTPUT: decoded string

```
initilise p ← root
while not EOF do
   read next bit b
   if b = 0 then
      p ← left_child(p)
   else
      p ← left_child(p)
   end if
   if p is a leaf then
      output the symbol at the leaf
      p ← root
   end if
end while
```

4.2 Shannon-Fano approach

Shannon-Fano coding is similar to Huffman[2] coding and only differs in the way of constructing the binary tree. In the Shannon-Fano approach, a binary tree

[2]The Shannon-Fano method was proposed before Huffman coding and is named after the inventors Claude Shannon (Bell Laboratories) and Robert Fano (MIT).

is constructed in a 'top-down' manner.

Let us first review Figure 4.2 in Example 4.1. There the root is the whole alphabet of the symbols `(((SN)L)EI)(B(AT))`. The brackets record how the symbols have been combined. They also provide a way to construct the binary tree according to the brackets from the root to the leaves.

In each iteration, a node can be spilt into two halves, one corresponds to a left subtree, the other to the right subtree.

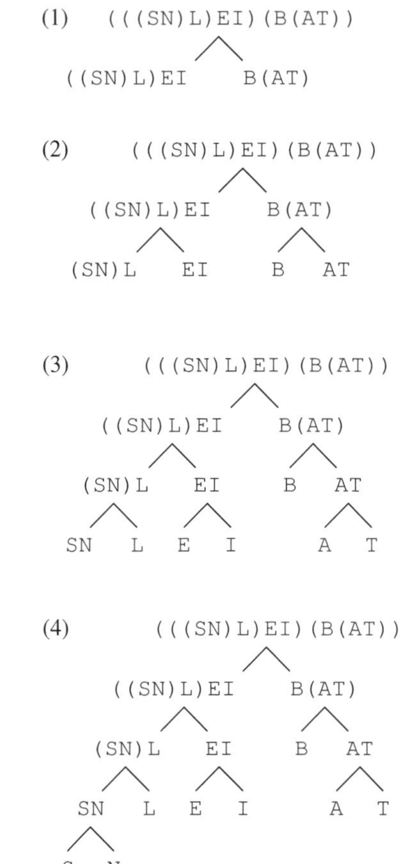

Figure 4.6: A binary tree constructed from the root

For example, from `(((SN)L)EI)(B(AT))`, we know the whole string can be divided into two halves: `((SN)L)EI` and `B(AT)`. Let the first half be the left child and the second the right child and we have a binary tree as in Figure 4.6(1). Next, each half can be divided into two halves again as in Figure 4.6(2). This division process continues until each half becomes a singleton symbol, as the whole binary tree in Figure 4.6(4).

The example suggests that, given an alphabet $\mathcal{S} = (s_1, s_2, \cdots, s_n)$, a binary tree can be constructed easily by dividing a string into two halves recursively. A 'middle' split point is required for each division. Suppose we know the probability distribution of the source $\mathcal{P} = (p_1, p_2, \cdots, p_n)$. The values of the probabilities of the symbols can then be used to find the middle point for each division. For example, we can on each iteration divide the symbol list into two halves with as balanced a weight as possible.

Example 4.3 *Consider a source* $\mathcal{S}$ = *(B, L, E, I, A, T, S, N) with frequencies (3, 2, 2, 1, 1, 1, 1, 1). We can divided the symbol list into two halves with the minimum difference between the sum of probabilities of the two halves.*

Figure 4.7 gives the process of the division. The vertical dash lines mark the division point for each segment of symbols. As we can see, after the first division (Figure 4.7(2)) the alphabet is split into two segments, `(B, L)` and `(E, I, A, T, S, N)`, with a minimum difference of frequency $|(2+3)-(2+1+1+1+1+1)| = 2$. After the next division, `(B, L)` is divided into `B` and `L`, and `(E, I, A, T, S, N)` is divided into `(E, I)` and `(A, T, S, N)` with a minimum frequency difference of $|3 - 2| = 1$ and $|(2 + 1) - (1 + 1 + 1 + 1)| = 1$ respectively. This process continues until step (5) (Figure 4.7(5)) where all the leaves are a single symbol.

This example shows the Shannon-Fano approach precisely. In fact, since our goal is to produce a prefix code instead of a binary tree, the process of labelling 0-1s can be embedded into the division process. For example, each time after a division, we can simply add a 0 to the codeword for the first half and a 1 to the second half (or a 1 to the first half and 0 to the second half). The codes derived in this way are called *Shannon-Fano codes.*

We now can derive the algorithm.

4.2.1 Shannon-Fano algorithm

Compression

Given a list of symbols, the algorithm involves the following steps:

1. Develop a frequency (or probability) table
2. Sort the table according to frequency (the most frequent one at the top)
3. Divide the table into two halves with similar frequency counts
4. Assign the upper half of the list a 0 and the lower half a 1
5. Recursively apply the step of division (3) and assignment (5) to the two halves, subdividing groups and adding bits to the codewords until each symbol has become a corresponding leaf on the tree.

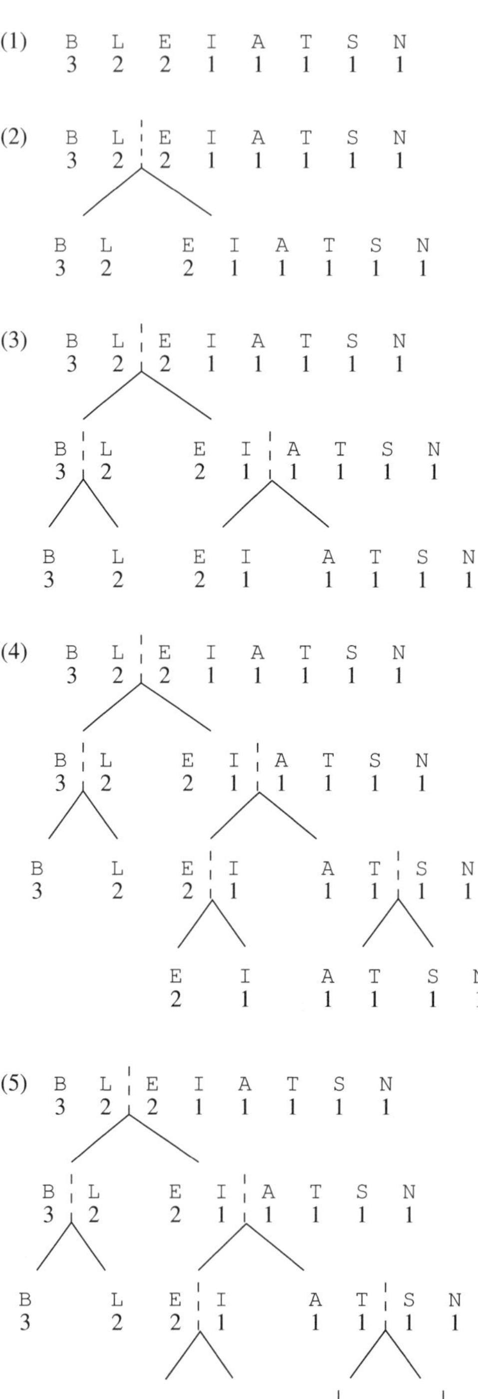

Figure 4.7: Building an equivalent Shannon-Fano tree

This can be further refined to Shannon-Fano(S) algorithm.

Algorithm 4.5 Shannon-Fano encoding ideas

```
if S consists of only two symbols then
    add 0 to the codeword of the first symbol and 1 to the second
else
    if S has more than three symbols then
        divide S into 2 subsequence S1 and S2 with minimum probability difference
        add a 0 to extend the codeword for each symbol in S1 and a 1 to those in S2
        Shannon-Fano(S1)
        Shannon-Fano(S2)
    end if
end if
```

Example 4.4 *Suppose the sorted frequency table below is drawn from a source. Derive the Shannon-Fano code.*

```
Symbol         A B C D E
Frequency     15 7 6 6 5
```

Solution

1. First division:

   (a) Divide the table into two halves so the sum of the frequencies of each half is as close as possible.

   ```
   Symbol          A B  | C D E
   Frequency      15 7  | 6 6 5
   sum(s)         (22)  | (17)
   ```

   (b) Assign one bit of the symbol (e.g. upper group 0s and the lower 1s).

   ```
   Symbol          A B  | C D E
   Frequency      15 7  | 6 6 5
   sum(s)         (22)  | (17)
   codewords       0    | 1
   ```

2. Second division:
   Repeat the above recursively to each group.

   ```
   Symbol          A | B,     C | D E
   Frequency      15 | 7,     6 | 6 5
   sum(s)        (15)|(7),  (6)|(11)
   codewords      00 | 01,   10 | 11
   ```

3. Third division:

```
Symbol        A, B, C,       D | E
Frequency    15, 7, 6,       6 | 5
codewords    00,01,10,     110 | 111
```

4. So we have the following code (consisting of five codewords) when the recursive process ends:

```
A    B    C    D     E
00   01   10   110   111
```

Decompression

Decompression takes the same approach as Huffman decoding. We leave it as an exercise. We have derived Huffman coding and Shannon-Fano coding algorithms. Before implementing the algorithms, we want to know the advantages and disadvantages of these algorithms.

Saving percentage

Saving percentage is an instance-based measure.

Consider Figure 4.3 in Example 4.1 again. Given the alphabet (`B, L, E, I, A, T, S, N`) with the frequency table (3, 2, 2, 1, 1, 1, 1, 1), the Huffman code (10, 001, 010, 011, 110, 111, 0000, 0001) is derived with lengths (2, 3, 3, 3, 3, 3, 4, 4) respectively. The total number of bits required by the source `BILLBEATSBEN` is $2 \times 3 + 3 \times 2 \times 2 + 3 \times 1 \times 3 + 4 \times 1 \times 2 = 35$.

Comparing this to the use of 8 bit ASCII or EBCDIC coding, the source `BILLBEATSBEN` requires a total of $8 \times 12 = 96$ bits.

Huffman	ASCII/EBCDIC	Saving bits	Percentage
35	96	61	63.5%
		$96 - 35 = 61$	$61/96 = 63.5\%$

The saving percentage seems impressive. However, for small alphabets like the one in this example, other coding methods can also be very effective. For example, if we know the alphabet for a source is always as small as eight symbols, a 3 bit fixed length code can be used as three bits can represent $2^3 = 8$ distinctive codewords. Comparing to a 3 bit fixed length code where the source requires $3 \times 12 = 36$ bits, Huffman coding achieves approximately 2.7% saving percentage.

Huffman	ASCII/EBCDIC	Saving bits	Percentage
35	36	1	2.7%
		$36 - 35 = 1$	$1/36 = 2.7\%$

Is the Huffman code in this instance the best code? To answer this question, we only need to compute the entropy of the source.

Entropy

We first convert the frequencies of the alphabet to probabilities (1/4, 1/6, 1/6, 1/12, 1/12, 1/12, 1/12, 1/12). Then compute the entropy:

$$H = \frac{1}{4}\log_2 4 + 2(\frac{1}{6}\log_2 6) + 5(\frac{1}{12}\log_2 12) \approx 2.855 \text{ bits}$$

We then compute the average length of the Huffman code with lengths (2, 3, 3, 3, 3, 3, 4, 4):

$$\bar{l} = 2 \times \frac{1}{4} + 2 \times 3 \times \frac{1}{6} + 3 \times 3 \times \frac{1}{12} + 2 \times 4 \times \frac{1}{12} \approx 4.416 \text{ bits}$$

The difference between the entropy and the average length of the code represents, in this instance, the room for any further improvement on compression. This means Huffman codes are not always optimal.

If the performance of a compression algorithm such as Huffman coding depends on the source, under what condition on the source does the Huffman compression algorithms perform the best? In other words, under what source condition are the Huffman codes optimal?

4.3 Optimal Huffman codes

Huffman codes are optimal when probabilities of the source symbols are all negative powers of two. Examples of a negative power of two are $\frac{1}{2}$, $\frac{1}{4}$, $\frac{1}{8}$, etc.

The conclusion can be drawn from the following justification.

Suppose that the lengths of the Huffman code are $\mathcal{L} = (l_1, l_2, \cdots, l_n)$ for a source $\mathcal{P} = (p_1, p_2, \cdots, p_n)$, where n is the size of the alphabet.

Using a variable length code to the symbols, l_j bits for s_j, the average length of the codewords is (in bits):

$$\bar{l} = \sum_{j=1}^{n} l_j p_j = l_1 p_1 + l_2 p_2 + \cdots + l_n p_n$$

The entropy of the source is:

$$H = \sum_{j=1}^{n} p_j \log \frac{1}{p_j} = p_1 \log \frac{1}{p_1} + p_2 \log \frac{1}{p_2} + \cdots + p_n \log \frac{1}{p_n}$$

As we know from Section 2.4.2, a code is optimal if the average length of the codewords equals the entropy of the source.

Let

$$\sum_{j=1}^{n} l_j p_j = \sum_{j=1}^{n} p_j \log_2 \frac{1}{p_j}$$

and notice

$$\sum_{j=1}^{n} l_j p_j = \sum_{j=1}^{n} p_j l_j$$

This equation holds if and only if $l_j = -\log_2 p_j$ for all $j = 1, 2, \cdots, n$, because l_j has to be an integer (in bits). Since the length l_j has to be an integer (in bits) for Huffman codes, $-\log_2 p_j$ has to be an integer, too. Of course, $-\log_2 p_j$ cannot be an integer unless p_j is a negative power of 2, for all $j = 1, 2, \cdots, n$.

In other words, this can only happen if all probabilities are negative powers of 2 in Huffman codes, for l_j has to be an integer (in bits). For example, for a source $\mathcal{P} = (\frac{1}{2}, \frac{1}{4}, \frac{1}{8}, \frac{1}{8})$, Huffman codes for the source can be optimal.

4.4 Implementation efficiency

The Huffman algorithms described earlier require a list of symbolic items to be maintained on each iteration in descending order of probabilities. The search for a right place for the newly combined item requires $O(n^2)$ time overall in the worst case, where n is the number of symbols in the alphabet.

One way to improve the time efficiency is to modify the encoding algorithm slightly as follows:

1. Maintain two probability (or frequency) lists: one (L_s) contains the original symbols in descending order of probability. The other (L_c), initially empty, is built to contain the 'combined items' only.
2. A new combined item is always placed to the *front* of list L_c. This can be achieved in $O(1)$ worst time, since there is no need for searching for a right place.

The next two items to be combined are the smallest items, the same as the two if it were in a normal Huffman encoding algorithm. However, both L_s and L_c have to be taken into consideration in the two-list approach. The two items for the next combination may be chosen among the two last symbols in L_s, two combined symbol items in L_c, or one combined symbol item and one singleton symbol in L_c and L_s respectively, whichever items have the least weight (probability).

The following example shows how this efficient approach works step by step.

Example 4.5 *Consider the source alphabet (`A, B, C, D, E, F, G, H, I, J`) and the probabilities (in %) 19, 17, 15, 13, 11, 9, 7, 5, 3, 1. Show how to construct a Huffman tree efficiently by maintaining two lists.*

In what follows, `Ls` represents the singleton list L_s, `P` the probabilities of the symbols and `Lc` the list L_c for combined symbol items:

1.
```
Ls: A   B   C   D   E   F  G  H  I  J
P:  19  17  15  13  11  9  7  5  3  1
Lc: Empty
```

```
2. Ls: A   B   C   D   E   F   G   H
   P:  19  17  15  13  11  9   7   5
   Lc: (IJ)
   P:  4

3. Ls: A   B   C   D   E   F   G
   P:  19  17  15  13  11  9   7
   Lc: (H (IJ))
   P:  9

4. Ls: A   B   C   D   E
   P:  19  17  15  13  11
   Lc: (FG) (H (IJ)))
   P:  16    9

5. Ls: A   B   C   D
   P:  19  17  15  13
   Lc: (E (H (IJ))) (FG)
   P:  20            16

6. Ls: A   B
   P:  19  17
   Lc: (CD) (E (H (IJ))) (FG)
   P:  28    20            16

7. Ls: A
   P:  19
   Lc: (B (FG)) (CD) (E (H (IJ)))
   P:  33        28   20

8. Ls: Empty
   Lc: (A (E (H (IJ)))) (B (FG)) (CD)
   P:  39                33       28
   Li: Empty

9. Lc: ((B (FG)) (CD)) (A (E (H (IJ))))
   P:  61               39
   Ls: Empty

10. Lc: (((B (FG)) (CD)) (A (E (H (IJ)))))
    P:  100
    Ls: Empty
```

Construct the binary tree recursively from the root:

```
1.        ((B(FG))(CD))(A(E(H(IJ))))
                0 /          \ 1
          (B(FG))(CD)    A(E(H(IJ)))
```

2.

```
        ((B(FG))(CD))(A(E(H(IJ))))
             0                 1

   (B(FG))(CD)     A(E(H(IJ)))
      0      1     0       1

   B(FG)    CD     A    E(H(IJ))
```

3.

```
        ((B(FG))(CD))(A(E(H(IJ))))
             0                 1

   (B(FG))(CD)          A(E(H(IJ)))
      0      1          0        1

   B(FG)      CD        A     E(H(IJ))
   0   1     0  1              0    1

   B   FG    C   D             E   H(IJ)
```

4.

```
        ((B(FG))(CD))(A(E(H (IJ))))
             0                  1

   (B(FG))(CD)          A(E(H(IJ)))
      0      1          0        1

   B(FG)      CD        A     E(H(IJ))
   0   1     0  1              0    1

   B          C   D            E
      FG                          H(IJ)
     0  1                         0   1

     F  G                         H   IJ
```

5.

```
        ((B(FG))(CD))(A(E(H (IJ))))
             0                  1

   (B(FG))(CD)          A(E(H(IJ)))
      0      1          0        1

   B(FG)      CD        A     E(H(IJ))
   0   1     0  1              0    1

   B          C   D            E
      FG                          H(IJ)
     0  1                         0   1

     F  G                         H
                                      IJ
                                     0  1

                                     I  J
```

So the code is:

```
B 000   F 0010   G 0011   C 010    D 011    A 10
E 110   H 1110   I 11110   J 11111
```

Observation

1. Huffman or Shannon-Fano codes are *prefix codes* (Section 2.3.3) which are uniquely decodable.
2. There may be a number of Huffman codes, for two reasons:
   (a) There are two ways to assign a 0 or 1 to an edge of the tree. In Figure 4.1, we have chosen to assign 0 to the left edge and 1 to the right. However, it is possible to assign 0 to the right and 1 to the left. This would make no difference to the compression ratio.
   (b) There are a number of different ways to insert a combined item into the frequency (or probability) table. This leads to different binary trees. We have chosen in the same example to:
      i. make the item at the higher position the left child
      ii. insert the combined item on the frequency table at the highest possible position.
3. For a canonical minimum-variance code, the differences among the lengths of the codewords turn out to be the minimum possible.
4. The frequency table can be replaced by a probability table. In fact, it can be replaced by any approximate statistical data at the cost of losing some compression ratio. For example, we can apply a probability table derived from a typical text file in English to any source data.
5. When the alphabet is small, a fixed length (less than 8 bits) code can also be used to save bits.

Example 4.6 *If the size of the alphabet set is smaller than or equal to 32, we can use 5 bits to encode each character. This would give a saving percentage of*

$$\frac{8 \times 32 - 5 \times 32}{8 \times 32} = 37.5\%$$

6. Huffman codes are fragile for decoding: the entire file could be corrupted even if there is a 1 bit error.
7. The average codeword length of the Huffman code for a source is greater and equal to the entropy of the source and less than the entropy plus 1 (Theorem 2.2).

4.5 Extended Huffman coding

One problem with Huffman codes is that they meet the entropy bound only when all probabilities are powers of 2. What would happen if the alphabet is binary, e.g. $\mathcal{S} = (\texttt{a}, \texttt{b})$? The only optimal case[3] is when $\mathcal{P} = (p_a, p_b)$, $p_a = 1/2$ and $p_b = 1/2$. Hence, Huffman codes can be bad.

Example 4.7 *Consider a situation when $p_a = 0.8$ and $p_b = 0.2$.*

Solution Since Huffman coding needs to use 1 bit per symbol at least, to encode the input, the Huffman codewords are 1 bit per symbol on average:

$$\bar{l} = 1 \times 0.8 + 1 \times 0.2 = 1 \text{ bit.}$$

However, the entropy of the distribution is

$$H(\mathcal{P}) = -(0.8 \log_2 0.8 + 0.2 \log_2 0.2) = 0.72 \text{ bit.}$$

The efficiency of the code is

$$\frac{H(\mathcal{P})}{\bar{l}} = \frac{0.72}{1} = 72\%$$

This gives a gap of $1 - 0.72 = 0.28$ bit. The performance of the Huffman encoding algorithm is, therefore, $0.28/1 = 28\%$ worse than optimal in this case.

The idea of extended Huffman coding is to encode a sequence of source symbols instead of individual symbols. The alphabet size of the source is *artificially* increased in order to improve the code efficiency. For example, instead of assigning a codeword to every individual symbol for a source alphabet, we derive a codeword for every two symbols.

The following example shows how to achieve this:

Example 4.8 *Create a new alphabet $\mathcal{S}' = (\texttt{aa}, \texttt{ab}, \texttt{ba}, \texttt{bb})$ extended from $\mathcal{S} = (\texttt{a}, \texttt{b})$. Let aa be A, ab be B, ba be C and bb be D. We now have an extended alphabet $\mathcal{S}' = (A, B, C, D)$. Each symbol in the alphabet $\mathcal{S}'$ is a combination of two symbols from the original alphabet $\mathcal{S}$. The size of the alphabet $\mathcal{S}'$ increases to $2^2 = 4$.*

Suppose symbol 'a' or 'b' occurs independently. The probability distribution for $\mathcal{S}'$, the extended alphabet, can be calculated as below:

$$p_A = p_a \times p_a = 0.64$$

$$p_B = p_a \times p_b = 0.16$$

$$p_C = p_b \times p_a = 0.16$$

$$p_D = p_b \times p_b = 0.04$$

[3] Here we mean the average number of bits of a code equals the entropy.

We then follow the normal static Huffman encoding algorithm (Section 4.1.2) to derive the Huffman code for $\mathcal{S}$.

The canonical minimum-variance code for $\mathcal{S}'$ is (0, 11, 100, 101), for A, B, C, D respectively. The average length is 1.56 bits for two symbols.

The original output became $1.56/2 = 0.78$ bit per symbol. The efficiency of the code has been increased to $0.72/0.78 \approx 92\%$. This is only $(0.78 - 0.72)/0.78 \approx 8\%$ worse than optimal.

This method is supported by the following Shannon's fundamental theorem of discrete noiseless coding:

Theorem 4.1 *For a source $\mathcal{S}$ with entropy $H(\mathcal{S})$, it is possible to assign codewords to sequences of m letters of the source so that the prefix condition is satisfied and the average length $\bar{l}_m$ of the codewords per source symbol satisfies*

$$H(\mathcal{S}) \le \frac{\bar{l}_m}{m} < \frac{1}{m}$$

Summary

Statistical models and heuristic approach give rise to celebrating static Huffman and Shannon-Fano algorithms. Huffman algorithms take a *bottom-up* approach while Shannon-Fano *top-down*. Implementation issues make Huffman code more popular than Shannon-Fano's. Maintaining two tables may improve the efficiency of the Huffman encoding algorithm. However, Huffman codes can give bad compression performance when the alphabet is small and the probability distribution of a source is skewed. In this case, extending the small alphabet and encoding the source in small groups of symbols may improve the overall compression.

Learning outcomes

On completion of your studies in this chapter, you should be able to:

- describe Huffman coding and Shannon-Fano coding
- explain why it is not always easy to implement the Shannon-Fano algorithm
- demonstrate the encoding and decoding process of Huffman and Shannon-Fano coding with examples
- explain how to improve the implementation efficiency of the static Huffman encoding algorithm by maintaining two sorted probability lists
- illustrate some obvious weaknesses of Huffman coding
- describe how to narrow the gap between the *minimum average number of binary bits* of a code and the *entropy* using the extended Huffman encoding method.

Exercises

E4.1 Derive a Huffman code for the string AAABEDBBTGGG.

E4.2 Derive a Shannon-Fano code for the same string.

E4.3 Provide an example to show step by step how the Huffman decoding algorithm works.

E4.4 Provide a similar example for the Shannon-Fano decoding algorithm.

E4.5 Given an alphabet $\mathcal{S} = (A, B, C, D, E, F, G, H)$ of symbols with the probabilities 0.25, 0.2, 0.2, 0.18, 0.09, 0.05, 0.02, 0.01 respectively in the input, construct a canonical minimum-variance Huffman code for the symbols.

E4.6 Construct a canonical minimum-variance code for the alphabet A, B, C, D with probabilities 0.4, 0.3, 0.2 and 0.1 respectively. If the coded output is 101000001011, what was the input?

E4.7 Given an alphabet (a, b) with $p_a = 1/5$ and $p_b = 4/5$, derive a canonical minimum-variance Huffman code and compute:

(a) the expected average length of the Huffman code
(b) the entropy of the Huffman code.

E4.8 Following the Shannon-Fano code in Example 4.4, decode 0010001110100 step by step.

E4.9 Given a binary alphabet (X, Y) with $p_X = 0.8$ and $p_Y = 0.2$, derive a Huffman code and determine the average code length if we group three symbols at a time.

E4.10 Explain with an example how to improve the entropy of a code by grouping the alphabet.

E4.11 Derive step by step a canonical minimum-variance Huffman code for alphabet (A, B, C, D, E, F), given the probabilities below:

Symbol	Probability
A	0.3
B	0.2
C	0.2
D	0.1
E	0.1
F	0.1

Compare the *average length* of the Huffman code to the *optimal length* derived from the entropy distribution. Specify the unit of the codeword lengths used.

Hint: $\log_{10} 2 \approx 0.3$; $\log_{10} 0.3 \approx -0.52$; $\log_{10} 0.2 \approx -0.7$; $\log_{10} 0.1 = -1$.

Laboratory

L4.1 Derive and implement your own version of the Huffman algorithms in pseudocode.

L4.2 Construct two source files: the good and the bad. Explain what you mean by good and bad.

L4.3 Implement the Shannon-Fano algorithm.

L4.4 Comment on the difference between the Shannon-Fano and Huffman algorithms.

L4.5 Derive and implement algorithms for *canonical minimum-variance* Huffman codes.

L4.6 Implement the Huffman encoding and decoding algorithms using the *extended* coding method.

L4.7 Develop a computer program to demonstrate how the extended Huffman codes may improve the quality of compression.

Hint: The main idea here is to show the fact that the gap between the average length of a code and its entropy may be reduced using the extended Huffman coding method. If you have developed any programs in the previous laboratory sessions to compute the average length of a code and the entropy of a probability distribution, you may then simply integrate these into your program(s) in this section.

Assessment

S4.1 Explain how the implementation efficiency of a canonical minimum-variance Huffman coding algorithm can be improved by means of maintaining two frequency lists.

S4.2 Derive step by step a canonical minimum-variance Huffman code for alphabet (A, B, C, D, E, F) using the *efficient* implementation approach, given the probabilities that each character occurs in all messages are as follows:

Symbol	A	B	C	D	E	F
Probability	0.3	0.2	0.2	0.1	0.1	0.1

S4.3 Compute the average length of the Huffman code derived from the above question.

S4.4 Given $\mathcal{S} = (A, B, C, D, E, F, G, H)$ and the symbols' occurring probabilities 0.25, 0.2, 0.2, 0.18, 0.09, 0.05, 0.02, 0.01, construct a canonical minimum-variance Huffman code.

S4.5 Consider alphabet (A, B). Suppose the probability of A and B, p_A and p_B are 0.2 and 0.8 respectively. It has been claimed that even the best canonical minimum-variance Huffman coding is about 37% worse than its optimal binary code. Do you agree with this claim? If yes, demonstrate how this result can be derived step by step. If no, show your result with good reasons.

Hint: $\log_{10} 2 \approx 0.3$; $\log_{10} 0.8 \approx -0.1$; $\log_{10} 0.2 \approx -0.7$.

S4.6 For the above question:

(a) derive the alphabet that is expanded by grouping two symbols at a time

(b) derive the canonical Huffman code for this expanded alphabet

(c) compute the expected average length of the Huffman code.

Bibliography

[CLRS01] T.H. Cormen, C.E. Leiserson, R.L. Rivest, and C. Stein. *Introduction to Algorithms*. The MIT Press, 2nd edition, 2001.

[Huf52] D.A. Huffman. A method for the construction of minimum-redundancy codes. *Proceedings of the Institute of Radio Engineers*, 40(9):1098–1101, September 1952.

[Rub76] F. Rubin. Experiments in text file compression. *Communications of the ACM*, 19(11):617–623, November 1976.

Chapter 5

Adaptive Huffman coding

In static Huffman coding, the probability distribution remains unchanged during the process of encoding and decoding. A source file is likely to read only once for coding purposes to avoid expensive preprocessing such as reading the entire source. An alphabet and probability distribution is often applied based on the previous experience. Such an estimated model can compromise the compression quality substantially. The amount of the loss in compression quality depends very much on how much the probability distribution of the source differs from the estimated probability distribution.

Adaptive Huffman coding algorithms improve the compression ratio by applying to the model the statistics based on the source content seen from the immediate past. An alphabet and its frequency table are dynamically adjusted after reading each symbol during the process of compression or decompression. Compared to static Huffman coding, the adaptive model is much more close to the real situation of the source after initial steps.

The adaptive Huffman coding technique was developed based on Huffman coding, first by Newton Faller and Robert G. Gallager and then improved by Donald Knuth and Jeffrey S. Vitter in 1985–87. In this chapter, we focus on the ideas behind the adaptive Huffman algorithms rather than specific versions by any authors.

5.1 Adaptive approach

In the adaptive Huffman coding, an alphabet and frequencies of its symbols are collected and maintained dynamically according to the source file on each iteration. The Huffman tree is also updated based on the alphabet and frequencies dynamically. When the encoder and decoder are at different locations, both maintain an identical Huffman tree for each step independently. Therefore, there is no need transferring the Huffman tree.

During the compression process, the Huffman tree is updated each time after a symbol is read. The codeword(s) for the symbol is output immediately. For convenience of discussion, the frequency of each symbol is called the *weight* of the symbol to reflect the change of the frequency count at each stage.

The output of the adaptive Huffman encoding consists of Huffman codewords as well as fixed length codewords. For each input symbol, the output can be a Huffman codeword based on the Huffman tree in the previous step or a codeword of a fixed length code such as ASCII. Using a fixed length codeword as the output is necessary when a new symbol is read for the first time. In this case, the Huffman tree does not include the symbol yet. It is therefore reasonable to output the uncompressed version of the symbol. If the source file consists of ASCII, then the fixed length codeword would simply be the uncompressed version of the symbol.

In the encoding process, for example, the model outputs a codeword of a fixed length code such as ASCII code, if the input symbol has been seen for the *first* time. Otherwise, it outputs a Huffman codeword.

However, a mixture of the fixed length and variable length codewords can cause problems in the decoding process. The decoder needs to know whether the codeword should be decoded according to a Huffman tree or by a fixed length codeword before taking a right approach. A special symbol as a flag, therefore, is used to signal a switch from one type of codeword to another.

Let the current alphabet be the subset $\mathcal{S} = (\Uparrow, s_1, s_2, \cdots, s_n)$ of some alphabet α, and $g(s_i)$ be any fixed length codeword for s_i (e.g. ASCII code), $i = 1, 2, \cdots$. To indicate whether the output codeword is a fixed length or a variable length codeword, one special symbol $\Uparrow$ ($\notin \alpha$) is defined as a *flag* or a *shift* key and to be placed before the fixed length codeword for communication between the compressor and decompressor.

5.2 Compressor

The compression algorithm maintains a subset $\mathcal{S}$ of symbols of some alphabet α ($\mathcal{S} \subset \alpha$) that the system has seen so far. A Huffman code (i.e. the Huffman tree) for all the symbols in $\mathcal{S}$ is also maintained. Let the weight of $\Uparrow$ always be 0 and the weight of any other symbol in $\mathcal{S}$ be its frequency so far. For convenience, we represent the weight of each symbol by a number in round brackets. For example, A(1) means that symbol A has a weight of 1.

Initially, $\mathcal{S} = \{\Uparrow\}$ and the Huffman tree has the single node of symbol $\Uparrow$ (see Figure 5.1 step (0)). During the encoding process, the alphabet $\mathcal{S}$ grows in number of symbols each time a new symbol is read. The weight of a new symbol is always 1 and the weight of an existing symbol in $\mathcal{S}$ is increased by 1 when the symbol is read. The Huffman tree is used to assign codewords to the symbols in $\mathcal{S}$ and is updated after each output.

Let $h(s_i)$ be the current Huffman codeword for s_i and `SHIFT` for the special symbol $\Uparrow$ in the algorithms next.

The following example shows the idea of the adaptive Huffman coding.

Example 5.1 *Suppose that the source file is a string* `ABBBAC`. *Figure 5.1 shows states of each step of the adaptive Huffman encoding algorithm.*

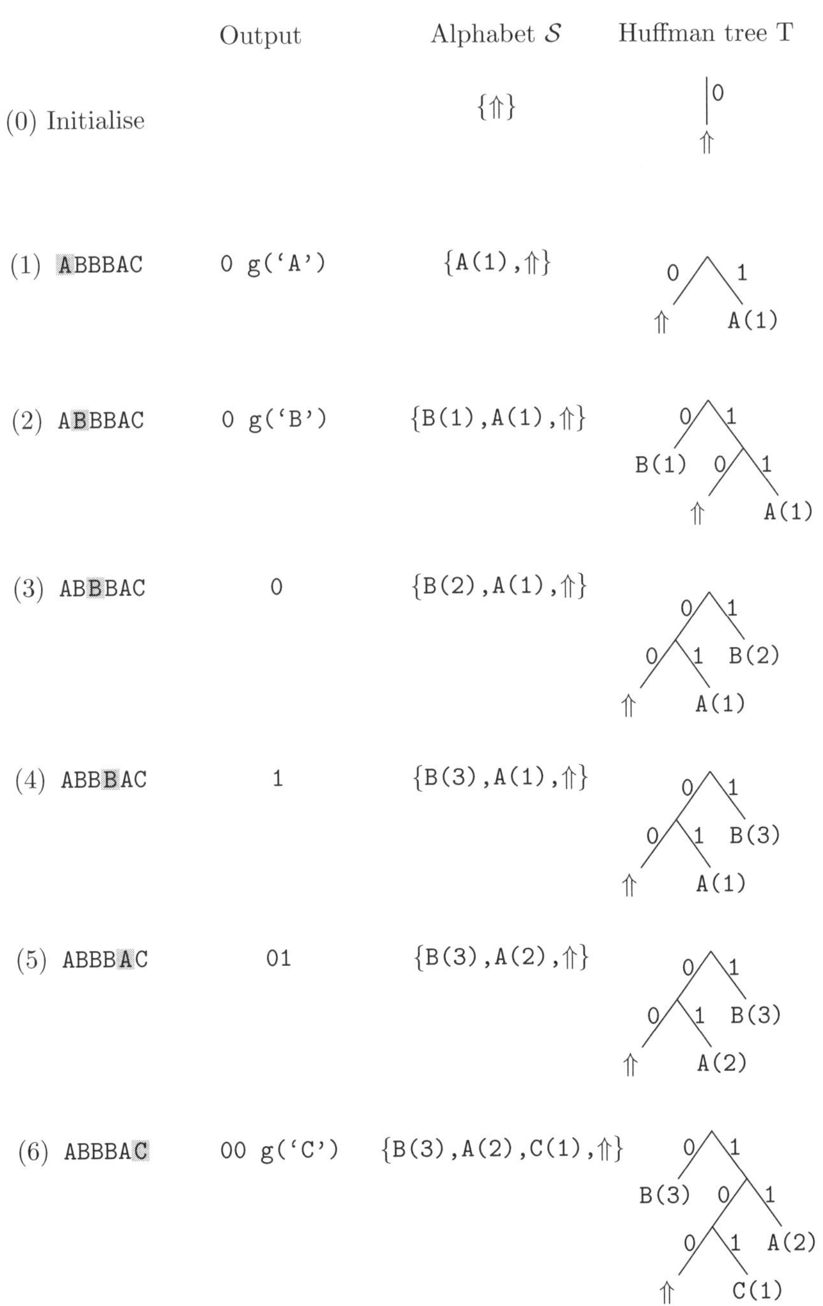

Figure 5.1: An example of adaptive Huffman encoding

As we can see from Figure 5.1, one symbol X is read from the source and is marked X in the input string. Alphabet $\mathcal{S}$ contains initially a single element ⇑ with a weight of zero. As each symbol X is read, the alphabet either grows in number of symbols or in weight of an existing symbol, but $\Uparrow$ remains a zero weight. The Huffman tree is updated on each step accordingly. The output on each step can be a Huffman codeword for $\Uparrow$ followed by a fixed length codeword $g(X)$ such as steps (1) and (2) if X is a new symbol, or a Huffman codeword for the input symbol such as steps (4) and (5) if X is an existing symbol in $\mathcal{S}$ already.

5.2.1 Encoding algorithm

From Example 5.1, we can derive the following main statements of the encoding algorithm for each iteration.

$s \leftarrow$ *next_symbol_in_text()*
if s has been seen before **then**
    output $h(s)$
else
    output h(SHIFT) followed by $g(s)$
end if
$T \leftarrow$ *update_tree(T)*

Adding the initial statement and a `while` statement for repeating the interaction, the following encoding algorithm shows how the compressor works: Here

Algorithm 5.1 Adaptive Huffman encoding

initialise the Huffman tree T containing the only node SHIFT.
while more characters remain **do**
    $s \leftarrow$ *next_symbol_in_text()*
    if s has been seen before **then**
        output $h(s)$
    else
        output h(SHIFT) followed by $g(s)$
    end if
    $T \leftarrow$ *update_tree(T)*
end while

next_symbol_in_text() is a function that reads one symbol from the input sequence. *update_tree()* is another function which does the following:

Algorithm 5.2 (Function) update_tree()

if s is not in $\mathcal{S}$ **then**
    add s to $\mathcal{S}$; $weight[s] \leftarrow 1$
else
    $weight[s] \leftarrow weight[s] + 1$
end if
recompute the Huffman tree for the new set of weights or symbols

5.3 Decompressor

The decompression algorithm also maintains a set of symbols $\mathcal{S}$ that the system has seen so far. The weight of $\Uparrow$ is always 0, and the weight of any other symbol is the frequency of occurrence so far in the decoded output. Initially, $\mathcal{S} = \{\Uparrow\}$.

Example 5.2 *Suppose the input at the decoding end is* `0g('A')0g('B')010100g('C')`*. Figure 5.2 shows states of each step of the adaptive Huffman decoding algorithm.*

As we can see from Figure 5.2, the decoder reads the compressed file bit by bit. The alphabet $\mathcal{S}$ grows in number of elements as in steps (1) and (2), or in weight of some existing symbols as in steps (3) and (4). The Huffman tree is identical to the Huffman tree for compression and updated on each step. The highlight path from the root to a leaf shows the corresponding input bits for each step.

Let us look at how the decompressor works step by step. In step (1), 0 is read. The decoder traces the only edge down from the root of the current Huffman tree (in step (0)) and finds the leaf is the special symbol $\Uparrow$. This indicates that what follows is a fixed length codeword. The decoder then reads the fixed length codeword and outputs the original symbol 'A'. The first-time-seen symbol 'A' with weight 1 is added to the alphabet $\mathcal{S}$ which then becomes $\{A(1), \Uparrow\}$. The Huffman tree is updated accordingly (in step (1)). In step (2), a 0 is read. The decoder traces the 0 edge in the current Huffman tree (in step (1)) and finds the leaf is the $\Uparrow$ again. So the next fixed length codeword g('B')[1] is read and B is output. The new symbol B(1) is added into the alphabet $\mathcal{S}$ and the Huffman tree is updated. In step (3), a 0 is read. One 0 edge is traced and leaf 'B' is reached in the current Huffman tree (in step (2)). So symbol B is output. The weight of B is increased by 1 and the Huffman tree is updated. In step (4), a 1 is read. One 1-edge is traced and leaf 'B' is output. The weight of B is increased by 1 and the Huffman tree is updated again. In step (5), a 0 and then a 1 are read before reaching a leaf A(1) in the current Huffman tree (in step (4)). So symbol A is output. The weight of A is increased by 1 in the alphabet and the Huffman tree is updated. In step (6), a 0 and then a 0 are read and a $\Uparrow$ is reached. So the next fixed length codeword g('C') is read. C is output and the new symbol C(1) is added into the alphabet and the Huffman tree is updated.

[1] For example, the decoder actually reads the next 8 bits if the g('B') is an 8 bit extended ASCII codeword for symbol B.

	Output	Alphabet $\mathcal{S}$	Huffman tree T
(0) Initialise		{⇑}	
(1) 0g('A')0g('B')010100g('C')			
0g('A')0g('B')010100g('C')	A	{A(1),⇑}	
(2) 0g('A')0g('B')010100g('C')			
0g('A')0g('B')010100g('C')	B	{B(1),A(1),⇑}	
(3) 0g('A')0g('B')010100g('C')	B	{B(2),A(1),⇑}	
(4) 0g('A')0g('B')010100g('C')	B	{B(3),A(1),⇑}	
(5) 0g('A')0g('B')010100g('C')			
0g('A')0g('B')010100g('C')	A	{B(3),A(2),⇑}	
(6) 0g('A')0g('B')010100g('C')			
0g('A')0g('B')010100g('C')			
0g('A')0g('B')010100g('C')	C	{B(3),A(2),C(1),⇑}	

Figure 5.2: An example of adaptive Huffman decoding

5.3.1 Decoding algorithm

The decoding algorithm can be summarised as below:

Algorithm 5.3 Adaptive Huffman decoding

```
initialise the Huffman tree T with single node SHIFT
while more bits remain do
    s ← huffman_next_sym()
    if s = SHIFT then
        s ← read_unencoded_sym()
    else
        output s
    end if
    T ← update_tree(T)
end while
```

The function *huffman_next_sym()* reads bits from the input until it reaches a leaf node and returns the symbol with which that leaf is labelled.

Algorithm 5.4 (Function) huffman_next_sym()

```
start at root of Huffman tree
while not reach a leaf do
    read_next_bit()
    traverse one edge down
end while
return the symbol of leaf reached
```

The function *read_unencoded_sym()* simply reads the next *unencoded* symbol from the input. For example, if the original encoding was an ASCII code, then it would read the next 8 bits (including the parity bit).

As in Section 5.2, the function *update_tree* does the following:

Algorithm 5.5 (Function) update_tree

```
if s is not in S then
    add s to S
else
    weight[s] ← weight[s] + 1
end if
recompute the Huffman tree for the new set of weights and symbols.
```

5.4 Disadvantages of Huffman algorithms

Adaptive Huffman coding has the advantage of requiring no preprocessing and the low overhead of using the uncompressed version of the symbols only at their first occurrence.

The algorithms can be applied to other types of files in addition to text files. The symbols can be objects or bytes in executable files.

Huffman coding, either static or adaptive, has two disadvantages that remain unsolved:

- **Disadvantage 1** It is not optimal unless *all* probabilities are negative powers of 2. This means that there is a gap between the average number of bits and the entropy in most cases.

  Recall the particularly bad situation for binary alphabets. Although by grouping symbols and extending the alphabet, one may come closer to the optimal, the blocking method requires a larger alphabet to be handled. Sometimes, extended Huffman coding is not that effective at all.

- **Disadvantage 2** Despite the availability of some clever methods for counting the frequency of each symbol reasonably quickly, it can be very *slow* when rebuilding the entire tree for each symbol. This is normally the case when the alphabet is big and the probability distributions change rapidly with each symbol.

Summary

Adaptive Huffman coding works on dynamic statistical models. The statistical models may be adopted to work more closely with the coders. The probability distribution is computed applying a frequency count and adjustment of alphabet after each new symbol being input into the coder. Two types of codes are used and a switch codeword is used to flag the alternative use of the codes.

Learning outcomes

On completion of your studies in this chapter, you should be able to:

- distinguish a static compression system from an adaptive one
- describe how adaptive Huffman coding algorithms work with examples
- identify implementation issues of Huffman coding algorithms
- explain the two problems (weaknesses) of the Huffman coding algorithms in general.

Exercises

E5.1 Explain, with an example, how adaptive Huffman coding works.

E5.2 Trace the adaptive Huffman coding algorithms to show how the following sequence of symbols is encoded and decoded:

```
aaabbcddbbccc
```

Laboratory

L5.1 Design and implement a simple version of the adaptive Huffman encoding algorithm.

L5.2 Design and implement a simple version of the adaptive Huffman decoding algorithm.

Assessment

S5.1 Describe briefly how each of the two classes of lossless compression algorithms, namely the *adaptive* and the *non-adaptive*, works in its model. Illustrate each with an appropriate example.

S5.2 Show how to encode the sequence below step by step using the adaptive Huffman coding algorithm.

```
abcbbdaaddd
```

Bibliography

[CH84] G.V. Cormack and R.N. Horspool. Algorithms for adaptive Huffman codes. *Information Processing Letters*, 18(3):159–165, March 1984.

[Fal73] N. Faller. An adaptive system for data compression. In *Record of the 7th Asilomar Conference on Circuits, Systems and Computers*, pages 593–597, Piscataway, NJ, 1973. IEEE Press.

[Gal78] R.G. Gallager. Variations on a theme by Huffman. *IEEE Transactions on Information Theory*, IT-24(6):668–674, November 1978.

[Knu85] D.E. Knuth. Dynamic Huffman coding. *Journal of Algorithms*, 6(2):163–180, June 1985.

[Vit87] J.S. Vitter. Design and analysis of dynamic Huffman codes. *Journal of the ACM*, 34(4):825–845, October 1987.

[Vit89] J.S. Vitter. Algorithm 673: Dynamic Huffman coding. *ACM Transactions on Mathematical Software*, 15(2):158–167, June 1989.

Chapter 6

Arithmetic coding

Arithmetic coding is important historically because it was at the time the most successful alternative to Huffman coding after a gap of 25 years. It is superior in performance to the Huffman coding especially when the alphabet is fairly small. The arithmetic method extended the early coding work by Shannon, Fano and Elias and was developed largely by Pasco (1976), Rissanene (1976, 1984), and Langdon (1984). It bypasses the idea of replacing every single input symbol with a codeword. Instead, it encodes a stream of input symbols with a single fraction as the compressed output.

6.1 Probabilities and subintervals

The idea comes first from Shannon's observation in 1948 that messages N symbols long may be encoded by their cumulative probability. This can also be seen from the grouping of symbols in static Huffman coding where a sequence of symbols is assigned one Huffman codeword to achieve a better compression ratio. It is possible to accomplish the same task without explicitly extending the alphabet. The arithmetic method is based on the fact that the cumulative probability of a symbol sequence corresponds to a unique subinterval of the initial [0, 1], and an assumption that the alphabet is small. However, it took a long time to solve the so-called *precision problem* before arithmetic algorithms became useful.

We first look at the following example to show how the idea works.

Example 6.1 *Consider a binary source alphabet (A, B) with a probability distribution (0.2, 0.8).*

Figure 6.1(a) shows the initial interval [0, 1].

Suppose that an input string contains only 1 symbol. The current interval [0, 1)[1] can be divided into two subintervals according to the probability distri-

[1][0, 1) means all real numbers ≥ 0 and < 1.

bution (p_A, p_B), where a symbol A corresponds to the left subinterval and B to the right (Figure 6.1(b)).

Suppose that an input string contains only 2 symbols. The current interval can be divided into two subintervals further according to the probability distribution (p_A, p_B), where a symbol A corresponds to the left subinterval and B to the right. For instance, sequence BA corresponds to the highlighted interval in Figure 6.1(c).

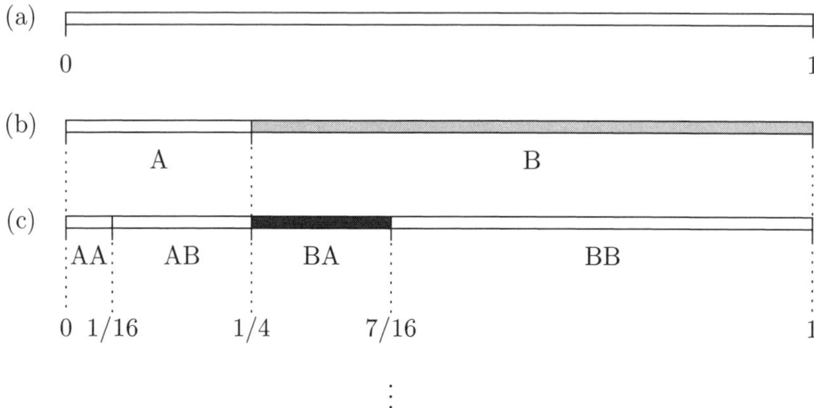

Figure 6.1: Encoding for two symbols

This is similar to an extended binary alphabet, where each string of certain length has a one-to-one relationship to a unique cumulative probability interval:

Sequence	Probability	Cumulative Probability	Interval
AA	$p_A \times p_A = 1/16$	1/16	[0, 1/16)
AB	$p_A \times p_B = 3/16$	1/4	[1/16, 1/4)
BA	$p_B \times p_A = 3/16$	1/4	[1/4, 7/16)
BB	$p_B \times p_B = 9/16$	7/16	[7/16, 1)

However, there is no need to extend the alphabet and compute every combined probability. Suppose that a string ABBBAABAAA of 10 symbols with probability distribution (0.2, 0.8) is to be encoded and the occurrence of each symbol is independent. Firstly, instead of extending the alphabet to one with 2^{10} combined elements, the cumulative probability of ABBBAABAAA and the corresponding interval can be computed easily as $p = 0.2 \times 0.8 \times 0.8 \times 0.8 \times 0.2 \times 0.2 \times 0.8 \times 0.2 \times 0.2 \times 0.2 = 2.6214 \times 10^{-5}$. The final result, 2.6214×10^{-5}, can be viewed as an adaptive and accumulate process of reading one symbol at a time and of including the probability of each symbol into its partial product at each iteration. Secondly, multiplying a real number at iteration i, say $p_i \in [0,\ 1)$, to a number x is equivalent to taking a proportion of the x value; the iteration process can be combined with the process of reading the input string. Once the ith symbol is read, the probability of the symbol p_i is multiplied immediately to update the current interval for the next round. A simple choice of the initial interval is [0, 1).

The output of arithmetic coding is essentially a fraction in the interval [0, 1). This single number can be uniquely decoded to create the exact stream of symbols that went into its construction.

The arithmetic coding algorithm overcomes the disadvantages of Huffman coding discussed in Section 5.4. It encodes *a sequence* of symbols at a time, instead of a single symbol. This may reduce the difference in value between the entropy and the average length. The algorithm does not output any codeword until only after seeing the *entire* input. This would be more efficient than the extended Huffman method by grouping symbols. The oversized alphabet is no longer an issue because only the probability of the input string is required.

6.2 Model and coders

The model for arithmetic coding is similar to the model for Huffman coding. It is based on an alphabet and the probability distribution of its symbols. Ideally, the probabilities are computed from the precise frequency counts of a source. This requires reading the entire source file before the encoding process. In practice, we can use an estimated or fixed approximate probability distribution with the cost of a slightly lower compression ratio.

Similar to Huffman coding, the model can be a static or dynamic one depending on whether the probability distribution is changed during a coding process. We discuss static arithmetic coding in this book unless otherwise stated.

We first introduce a simple version of the arithmetic coder for binary sources.

Suppose that both encoder and decoder know the length of the source sequence of symbols. Consider a simple case with a binary alphabet (A, B), and the probability that the next input symbol 'A' is p_A and 'B' is p_B.

Compression

Our essential goal is to assign a unique interval to each potential symbol sequence of a known length. After deriving a unique interval for a given symbol sequence, what we need to do within the interval is merely select a suitable decimal number as the arithmetic codeword.

The arithmetic encoder reads a sequence of source symbols one symbol at a time. Each time a new subinterval is derived according to the probability of the input symbol. This process, starting with initial interval [0, 1), is iterated until the end of the input symbol sequence. Then the arithmetic coder outputs a chosen *real number* within the final subinterval for the entire input symbol sequence.

Example 6.2 *Suppose $p_A = 1/4$, and $p_B = 3/4$; and the symbol generation is 'memoryless'.*[2] *Show the ideas of the simple version of arithmetic coding for an input sequence containing 1. a single symbol and 2. two symbols.*

Solution

1. Since p_A is 1/4, we first divide the interval [0, 1) into two subintervals of which the size is proportional to the probability of each symbol, as shown in Figure 6.2, using the convention that a real number within $[0, 1/4)$ represents A and a real number from $[1/4, 1)$ represents B. Let A be 0.0 and B as 0.5.

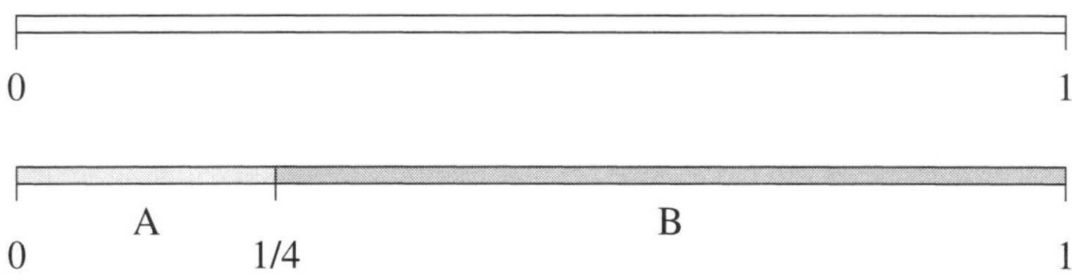

Figure 6.2: Encoding for one symbol

2. Extending this to two symbols, we can see that $p_{AA} = p_A \times p_A = 1/16$, $p_{AB} = p_A \times p_B = 3/16$, $p_{BA} = p_B \times p_A = 3/16$, and $p_{BB} = p_B \times p_B = 9/16$. We divide each of the two subintervals further into two sub-subintervals of which the size is again proportional to the probability of each symbol, as shown in Figure 6.3.

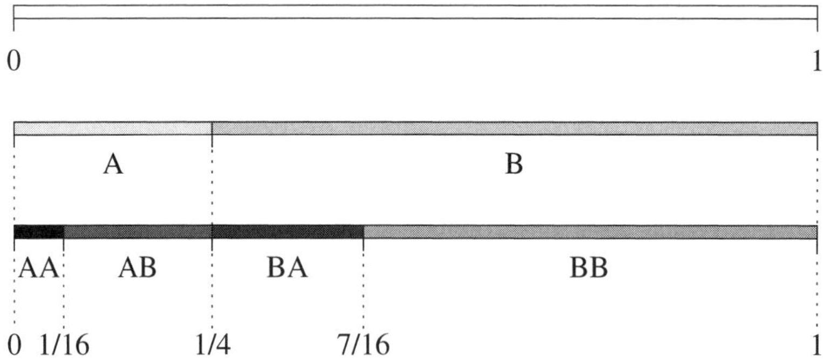

Figure 6.3: Encoding for two symbols

For example, when a symbol 'A' is read, the new interval can be derived from the original interval $(1 - 0)$ multiplied by p_A, i.e. $(1 - 0) \times 1/4$. The new interval becomes $[0, 1/4)$.

[2] By memoryless, we mean that the probability of the symbols are independent to each other.

If the next symbol is a 'B', the interval $[0, 1/4)$ will be divided further according to the probability of p_{AA} and p_{AB}, which are $p_A \times p_A = 1/4 \times 1/4 = 1/16$ and $p_A \times p_B = 1/4 \times 3/4 = 3/16$ respectively. The new interval is therefore [1/16, 1/4) for string 'AB'.

We may then encode for the final iteration as follows:

if the input so far is '`AA`', i.e. input an '`A`' followed by another '`A`' **then**
  output a number within $[0, 1/16)$, i.e. $[0, 0.0625)$, (output 0, for example)
else if the input is '`AB`' **then**
  output a number within [1/16, 1/4], i.e. [0.0625, 0.25), (output 0.1, for example)
else if the input is '`BA`' **then**
  output a number between $1/4 = 0.25$ and $1/4 + 3/16 = 7/16 = 0.4375$, (output 0.3, for example)
else
  output a number between $7/16 = 0.4375$ and $7/16 + 9/16 = 1$, (output 0.5, for example).
end if

Decompression

For convenience, the input of our decoding algorithm is a decimal fraction and the output is a sequence of symbols. In the decoding process, the initial interval is also [0, 1). We then determine on each iteration the subinterval according to which segment the input fraction falls. The subinterval that covers the fraction becomes the current interval for the following iteration and the corresponding symbol is output. This process repeats until the required number of symbols have been output.

Example 6.3 *Suppose that the probability distribution of a binary source is (1/4, 3/4). Outline the decoding process for 0.1, the codeword for a string of two symbols.*

Solution

1. Read 0.1.

   Since $p_A = 1/4 = 0.25$ and $p_A = 3/4 = 0.75$, the interval [0, 1) is divided into [0, 0.25) and [0.25, 1). The current interval is updated to [0, 0.25) for the next iteration, because $0 < 0.1 < 0.25$ (Figure 6.4(b)). So the corresponding symbol A is output.

2. Now the interval [0, 0.25) is divided according to the probabilities $p_A \times p_A = 1/4 \times 1/4 = 1/16$ and $p_A \times p_B = 1/4 \times 3/4 = 3/16$ and becomes

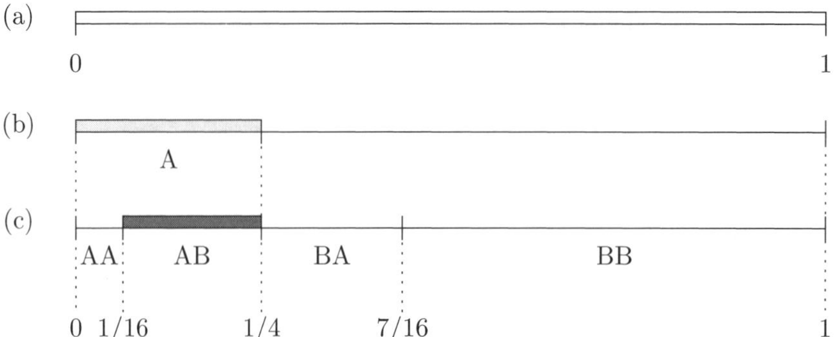

Figure 6.4: Decoding solution for two symbols

[0, 1/16) and [1/16, (1+3)/16)=[1/16, 1/4). The current interval is updated to [1/16, 1/4) for the next iteration, because $1/16 < 0.1 < 1/4$ (Figure 6.4(c)). The corresponding symbol B is output.

3. Since 2 is the length of the decoded string, we conclude that the decoded string is AB.

We now summarise the decoding ideas below:

```
currentInterval ← [0, 1)
while the number of output symbols is insufficient do
  divide currentInterval into [0, p_A) and [p_A, p_A + p_B) according to p_A
  and p_B
  read the codeword x
  if x in [0, p_A) then
    output 'A' and currentInterval ← [0, p_A)
  else
    output 'B' and currentInterval ← [p_A, p_A + p_B)
  end if
  update p_A and p_B
end while
```

Observation

From Example 6.2, you may have noticed the following facts:

1. When outputting uniquely the decimal number in the final interval, we could choose, in theory, *any* decimal that is in the range. However, various techniques are developed and applied to output a *short* codeword.

   For example, consider the final interval for string `AA` [0, 0.0625). We could have chosen, say, 0.05123456789 to encode `AA` but we do not, because 0 allows a 1-bit code, instead of an 11-bit one which would have been required for 0.05123456789.

2. It is possible to have variation in the arithmetic algorithms introduced here, because there are other equally good codewords. For example, `AB 0.2; BA 0.4; BB 0.6 (or 0.7, 0.8, 0.9)` would be fine if a unique final subinterval can be derived for every possible string from the initial interval [0, 1).

3. The codeword for `A` may be the same as the one for `AA` because of the possible overlap between their final intervals. Consider Example 6.2 where the final intervals for `A` and `AA` are [0, 1/4), and [0, 1/16) respectively, and the codeword for both `A` and `AA` can be a 0.

   However, this is acceptable because the encoder and the decoder are both assumed to know the length of the source sequence. Of course, the encoder does not need to know the length of the entire string in order to process it. It can stop the encoding process when an end-of-string sign is reached.

Similarly, we can extend this to a sequence with three symbols: we can easily work out the probability distribution (p_{AAA}, p_{AAB}, $\cdots$, p_{BBB}) and therefore decide easily the corresponding subinterval for a given string. In theory, this subinterval division process can be repeated as many times as the number of input symbols required.

The following table shows arithmetic encoding which can be applied to encode all length-3 strings over the alphabet (A, B) with $p_A = 1/4$ and $p_B = 3/4$.

Seq	Prob	Interval (fraction)	Interval (decimal)	Output	$\log_{10} p$
AAA	1/64	[0, 1/64)	[0, 0.015625)	0	1.81
AAB	3/64	[1/64, 4/64)	[0.015625, 0.0625)	0.02	1.33
ABA	3/64	[4/64, 7/64)	[0.0625, 0.109375)	0.1	1.33
ABB	9/64	[7/64, 16/64)	[0.109375, 0.25)	0.2	0.85
BAA	3/64	[16/64, 19/64)	[0.25, 0.296875)	0.25	1.33
BAB	9/64	[19/64, 28/64)	[0.296875, 0.4375)	0.3	0.85
BBA	9/64	[28/64, 37/64)	[0.4375, 0.578125)	0.5	0.85
BBB	27/64	[37/64, 1)	[0.578125, 1)	0.6	0.37

6.3 Simple case

From previous examples, we know that the encoding process is essentially a process to derive the final interval. This is achieved by the following main steps.

Here the interval is represented by $[L, L+d)$, where variable L stores lowest value of the interval range and d stores the distance between the highest and lowest value of the interval.

1. Let the initial interval be [0, 1).
2. Repeat the following until the end of the input sequence:

(a) read the next symbol **s** in the input sequence

(b) divide the current interval into subintervals whose sizes are proportional to the symbols' probabilities

(c) update the subinterval for the sequence up to **s** the *new* current interval.

3. (When the end of the input string is reached), output a decimal number within the current interval.

6.3.1 Encoding

Let p_1 be the probability for s_1 and p_2 be the probability for s_2, where $p_2 = 1 - p_1$. Let the current interval be $[L,\ L + d)$ at each stage, and the output fraction x satisfy $L \leq x < L + d$.

Initially, $L = 0$ and $d = 1$, this gives $[0,\ 1)$. If the next symbol could either be s_1 or s_2 with probability p_1 and p_2 respectively, then we assign the intervals

$$[L,\ \underline{L + d \times p_1}) \text{ and } [\underline{L + d \times p_1},\ \underline{L + d \times p_1} + d \times p_2)$$

and select the appropriate one. Note: $L+d\times p_1+d\times p_2 = L+d\times(p_1+p_2) = L+d$.

Algorithm 6.1 Encoding for binary source

```
1: L ← 0 and d ← 1
2: read next symbol
3: if next symbol is s1 then
4:    leave L unchanged and d ← d × p1
5: else
6:    L ← L + d × p1; d ← d × p2
7: end if
8: if no more symbols left then
9:    output a fraction from [L, L + d)
10: else
11:    go to step 2
12: end if
```

Example 6.4 *For the example earlier, let $s_1 = A$ and $s_2 = B$ with probabilities $p_1 = 1/4$ and $p_2 = 3/4$. We encode ABA as follows:*

- $L = 0$ *and* $d = 1$
- *read symbol A, leave* $L = 0$ *and set* $d = 1/4$
- *read symbol B, set* $L = 1/16$ *and* $d = 3/16$
- *read symbol A, leave* $L = 1/16$ *and set* $d = 3/64$
- *Done, so choose a decimal number* $\geq L = 1/16 = 0.0625$ *but* $< L + d = 0.109375$*, for example, choose 0.1 and output 1.*

6.3.2 Decoding

The decoding process is the inverse of the encoding process.

Let p_1 be the probability for s_1 and p_2 be the probability for s_2, where $p_2 = 1 - p_1$. Given a codeword x in [0, 1) and the length of the source string, the source can be decoded as below:

Algorithm 6.2 Decoding for binary source

```
1:  L ← 0 and d ← 1
2:  read x
3:  if x is a member of [L, L + d × p1) then
4:      output s1; leave L unchanged; d ← d × p1
5:  else
6:      output s2; L ← L + d × p1; d ← d × p2
7:  end if
8:  if numberOfDecodedSymbols < requiredNumberOfSymbols then
9:      go to step 2
10: end if
```

Observation

1. In the algorithm and examples that we have discussed so far, a decimal system (base 10) is used for (description) convenience. The average length of the code is

   $1/64 + 6/64 + 3/64 + 9/64 + 6/64 + 9/64 + 9/64 + 27/64 = 70/64 = 1.09$ digits

   The information theoretic bound, i.e. the entropy, is

   $$0.244 \times 3 = 0.73 \text{ digits}$$

   As a contrast, the average length of a static Huffman code would have required three digits in this case. This shows the strength of the arithmetic coding on small alphabets.

2. It is not difficult to modify the algorithms in this section slightly to output a sequence of 0s and 1s instead of a decimal. We leave this to the reader as an exercise.

6.4 General case

Let $S = (s_1, s_2, \cdots, s_n)$ be the alphabet of a source with an associated probability distribution of occurrence $P = (p_1, p_2, \cdots, p_n)$. The subintervals for each iteration can be derived for every symbol according to these probabilities. For example, after the first iteration, the initial iteration [0, 1) can be divided into n intervals as below:

$$\begin{array}{c}[0,\ p_1)\\ [p_1,\ \underline{p_1}+p_2)\\ [p_1+p_2,\ \underline{p_1+p_2}+p_3)\\ \vdots\\ [p_1+p_2+\cdots+p_{n-1},\ \underline{p_1+p_2+\cdots+p_{n-1}}+p_n)\end{array}$$

where $p_1 + p_2 + \cdots + p_n = 1$ for the n symbols in the alphabet $s_1, s_2, \cdots, s_n$ respectively. If the first symbol read is the ith symbol s_i in the alphabet, then the left boundary of the subinterval `low` is the cumulative probability $P_i = p_1 + p_2 + \cdots + p_{i-1}$ and the right boundary `high` is `low` $+ p_i$.

The length of the interval `high` $-$ `low` can then be used to compute one of the intervals in the next iteration:

$$\begin{array}{c}[\texttt{low},\ \texttt{low}+(\texttt{high}-\texttt{low})P_1)\\ [\texttt{low+(high-low)}P_1,\ \texttt{low}+(\texttt{high}-\texttt{low})P_2)\\ \vdots\\ [\texttt{low}+(\texttt{high}-\texttt{low})P_{n-1},\ \texttt{low}+(\texttt{high}-\texttt{low})P_n)\end{array}$$

where the cumulative probability on the whole set is $P_n = p_1+p_2+\cdots+p_n = 1$.

This looks complicated but there is no need to compute all the subintervals except one which depends on the independent probability of the symbol read at that iteration. As we can see from the algorithm later, the variables `low`, `high`, and the cumulative probability P_i can be easily updated.

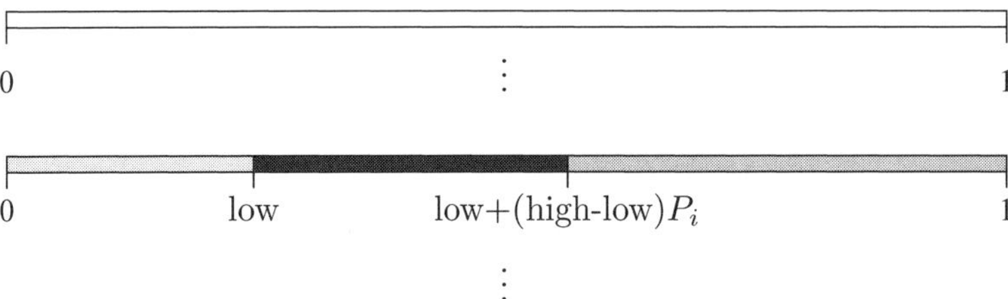

Figure 6.5: Encoding for a 3-symbol alphabet

Example 6.5 *Given three symbols A, B, C with probabilities 0.5, 0.3 and 0.2, the current allowed interval would be subdivided into three intervals according to the ratio 5:3:2. We would choose the new allowed interval among the three (Figure 6.5).*

Consider a source with an alphabet $(s_1, s_2, \cdots, s_n)$ and probability distribution $(p_1, p_2, \cdots, p_n)$. Suppose the length of the source is N.

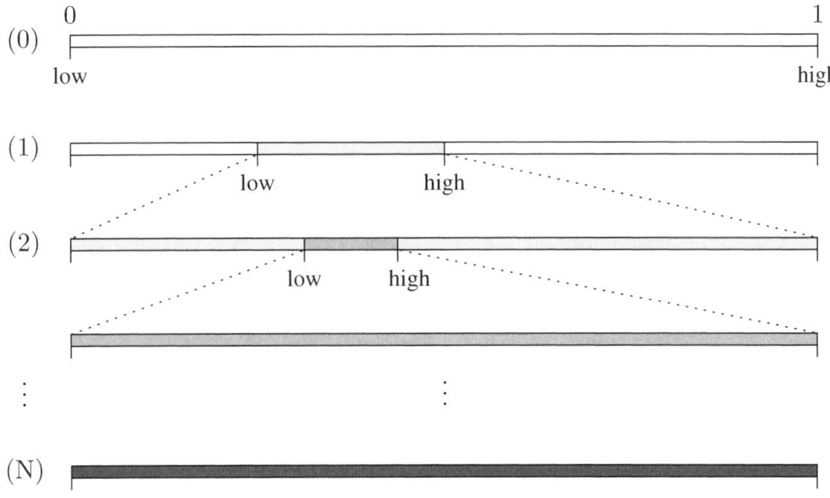

Figure 6.6: General case

The arithmetic algorithm for compression and decompression is given below: we use two variables `low` and `high` to define the arithmetic interval [`low, high`), where the interval can be easily changed by updating the values of `low` or `high` (Figure 6.6).

6.4.1 Compression algorithm

The algorithmic idea can be summarised as:

$currentInterval \leftarrow [0,\ 1)$
while not EOF **do**
  read a symbol s
  divide $currentInterval$ into subintervals according to the probability distribution P
  $currentInterval \leftarrow subinterval(s)$
end while
output a decimal from $currentInterval$

Let *low*, *high*, and *codeRange* be a real.

Algorithm 6.3 Arithmetic encoding

```
low ← 0.0
high ← 1.0
while there are still input symbols do
    get an input symbol s
    codeRange ← high − low
    high ← low + range × high_range(s)
    low ← low + range × low_range(s)
end while
output low
```

6.4.2 Decompression algorithm

The decompression ideas are:

```
currentInterval ← [0, 1);
for i ← 1; i ≤ SequenceLength; i ← i + 1 do
    divide currentInterval into subintervals according to
    the probability distribution P
    currentInterval ← subinterval(s)
    output symbol s corresponding to currentInterval
end for
```

Algorithm 6.4 Arithmetic decoding

```
get encoded number
repeat
    find symbol whose range covers the encoded number
    output the symbol
    range ← symbol_high_value − symbol_low_value
    subtract symbol_low_value from encoded-number
    divide encoded_number by range
until no more symbols
```

6.4.3 Unique decodability

We have shown in the previous section how it is possible to divide an interval starting from [0, 1) according to the probability of an input symbol each time and how to assign the final interval to the entire given symbol sequence of a certain length.

As we can see from the example and Figure 6.5, in theory, there is no overlap among these subintervals for all possible symbol sequences. It is this 'overlap-free' fact that makes arithmetic codes uniquely decodable.

6.4.4 Advantages of using fractions

To show this, we first look at some statistical properties of blocking symbols which is called *asymptotic equipartition property.* The result can be proved mathematically but is beyond the scope of this book. Nevertheless, one of the results is of interest. It explains why, by using fractions, arithmetic coding can achieve better compression results than Huffman coding.

Let $c_i = s_1 s_2 \cdots s_n$ be a grouped sequence of symbols from an alphabet of a source $\mathcal{S}$ with a probability distribution P. As we have seen before, if the source is so-called 'memoryless', the probability of the sequence of symbols is the product of the probabilities of each symbol in the sequence. That is, $p_c = p_1 p_2 \cdots p_n$. Hence,

$$-\frac{1}{n}\log_2(p_c) = -\frac{1}{n}\sum_{i=1}^{n}\log_2(p_i)$$

When n is big,

$$-\frac{1}{n}\sum_{i=1}^{n}\log_2(p_i) \approx -\frac{1}{n}\sum_{i=1}^{|S|}\log_2(p_i) = H(S)$$

So, the logarithm of the probability of the typical n combined symbols can be derived approximately from the n times of the entropy of the source,[3] i.e.

$$\sum_{i=1}^{n}\log_2(p_i) \approx nH(S)$$

The value $nH(S)$ is in general a fraction instead of an integer. That is why a decimal number of bits codeword such as in arithmetic is better than only allowing an integer number of bits, e.g. as in Huffman code.

6.4.5 Renormalisation

The number of x's digits in the decoding algorithm in Section 6.3.2 can be quite large. On most computers, the width of the current interval would become zero rapidly (say after a few hundred symbols or so). This is an example of the so-called *precision problem* for arithmetic coding, which is an interesting research subject in its own right.

Renormalisation is a technique for dealing with an implementation problem. The idea is to stop d and L becoming zero rapidly by resetting the intervals.

[3]For a more general situation where the symbols are not necessarily memoryless, the above equation can be used by just replacing the entropy by the *entropy rate.*

Example 6.6 *Consider the arithmetic encoding for* `BAABAA`*.*

After `BAA` *the interval is [0.25, 0.296875), and we know that the first digit of the output must be 25. So the decoder outputs 25 and resets the interval to [0.5, 0.96875); both become two to three times as big. This stops d and L going to zero.*

Summary

Arithmetic coding is a popular compression algorithm after Huffman coding and it is particularly useful for a relatively small and skewed alphabet. We only discuss the static approach here. In theory, an arithmetic coding algorithm encodes an entire file as a sequence of symbols into a single decimal number. The input symbols are processed one at each iteration. The initial interval [0, 1) (or [0, 1]) is successively divided into subintervals on each iteration according to the probability distribution. The subinterval that corresponds to the input symbol is selected for next iteration. The interval derived at the end of this division process is used to decide the codeword for the entire sequence of symbols. Unfortunately, implementation often encounters difficulties due to the constraints of computer precision.

Learning outcomes

On completion of your studies in this chapter, you should be able to:

- explain the main ideas of arithmetic coding
- describe, with an example of a small alphabet (A, B), how arithmetic encoding and decoding algorithms work
- discuss the advantages of arithmetic coding compared with Huffman coding
- explain the main problems in the implementation of arithmetic coding.

Exercises

E6.1 What are the advantages of arithmetic coding compared with the disadvantages of Huffman coding? You may use a simple version of the encoding algorithm such as Algorithm 6.1 in the discussion.

E6.2 Demonstrate how to encode a sequence of five symbols, namely `BABAB` from the alphabet (A, B), using the arithmetic coding algorithm if $p_A = 1/5$ and $p_B = 4/5$.

E6.3 Using the coding in the previous question as an example, explain how the arithmetic decoding algorithm works, for example, how to get the original `BABAB` back from the compressed result.

E6.4 Show how to encode a sequence of five symbols from (A, B), namely `ABABB` using arithmetic coding if $p_A = 1/5$ and $p_B = 4/5$. Would it be possible to have a coded output derived from a decimal value 0.24? Describe how the decoding algorithm works.

E6.5 A sequence of four symbols from (A, B) was encoded using arithmetic coding. Assume $p_A = 1/5$ and $p_B = 4/5$. If the coded output is 0.24, derive the decoded output step by step.

Laboratory

L6.1 Implement a simple version of the arithmetic encoding algorithm for a binary alphabet, e.g (A, B).

L6.2 Implement a simple version of the arithmetic decoding algorithm for the same binary alphabet (A, B).

L6.3 Implement a simple version of the arithmetic encoding and decoding algorithms for an alphabet of size n.

Assessment

S6.1 Describe briefly how arithmetic coding gets around the problems of Huffman coding. You may use a simple version of the encoding algorithm such as Algorithm 6.1 for discussion.

S6.2 Show how arithmetic coding can be applied to compress all length-4 strings over the alphabet (A, B). Suppose $p_A = 1/4$ and $p_B = 3/4$. You may like to summarise the code in the following table format:

Sequence	Probability	Interval fraction	Interval decimal	Output
?	?	?	?	?
⋮	⋮	⋮	⋮	⋮

Bibliography

[Gua80] M. Guauzzo. A general minimum-redundancy source-coding algorithm. *IEEE Transactions on Information Theory*, IT-26:15–25, January 1980.

[HV4a] P.G. Howard and J.S. Vitter. Arithmetic coding for data compression. *Proceedings of the IEEE*, 82(6):857–865, June 1994a.

[Lan84] G.G. Langdon. An introduction to arithmetic coding. *IBM Journal of Research and Development*, 28(2):135–149, March 1984.

[MNW98] A. Moffat, R.M. Neal, and I.H. Witten. Arithmetic coding revisited. *ACM Transactions on Information Systems*, 16(3):256–294, July 1998.

[Pas76] R. Pasco. *Source coding algorithms for fast data compression.* PhD thesis, Department of Electrical Engineering, Stanford University, 1976.

[PM88] W.B. Pennebaker and J.L. Mitchell. Probability estimation for the Q-Coder. *IBM Journal of Research and Development*, 32(6):737–752, November 1988.

[Ris76] J.J. Rissanen. Generalized Kraft inequality and arithmetic coding. *IBM Journal of Research and Development*, 20:198–203, May 1976.

[RL79] J.J. Rissanen and G.G. Langdon. Arithmetic coding. *IBM Journal of Research and Development*, 23(2):149–162, March 1979.

[Rub79] F. Rubin. Arithmetic stream coding using fixed precision registers. *IEEE Transactions on Information Theory*, IT-25(6):672–675, November 1979.

[WNC87] I.H. Witten, R.M. Neal, and J.G. Cleary. Arithmetic coding for data compression. *Communications of the Association for Computing Machinery*, 30(6):520–540, June 1987.

Chapter 7

Dictionary-based compression

Arithmetic algorithms as well as Huffman algorithms are all based on a statistical model, namely an alphabet and the probability distribution of a source. The compression efficiency for a given source depends on the alphabet size and how close its probability distribution of the statistics is to those of the source. The coding method also affects the compression efficiency. For a variable length code, the lengths of the codewords have to satisfy the Kraft inequality in order to be uniquely decodable. This, in one way, provides theoretically guidance on how far a compression algorithm can go; in another, it restricts the performance of these compression algorithms.

In this chapter, we look at a set of algorithms based on a dictionary instead of a statistical model. The dictionary is used to store the string patterns seen before and the indexes are used to encode the repeated patterns. The dictionary appears in either an explicit or an implicit form as we shall see later.

Dictionary compression approaches apply various techniques that incorporate the structure in the data in order to achieve a better compression. The goal is to eliminate the redundancy of storing repetitive strings for words and phrases repeated within the text stream. The coder keeps a record of the most common words or phrases in a document called a dictionary and uses their indices in the dictionary as output tokens. Ideally, the tokens are much shorter in comparison with the words or phrases themselves and the words and phrases are frequently repeated in the document.

The encoder reads the input string, identifies those recurrent words, and outputs their indices in the dictionary. A new word is output in the uncompressed form and added into the dictionary as an new entry. The main operations involve the comparison of strings, dictionary maintenance and an efficient way of encoding.

Compressors and decompressors both maintain a dictionary by themselves. The dictionary-based algorithms are normally faster than entropy-based ones.

They process the input as a sequence of characters rather than as streams of bits.

The input to the compression algorithm is a stream of symbols and the output consists of a mixture of tokens and words in original form. When it outputs tokens, the coding system can be classified as working in *variable-to-fixed* fashion since, in the basic form, each string to be encoded is of different length but the codewords, i.e. the indices in the dictionary, are of the same length.

Dictionary-based approaches are adaptive[1] in nature because the dictionary is updated during the process of compression and decompression. The content of the dictionary varies according to the input sequence of the text to be compressed.

Dictionary approaches do not use any statistical model but rely upon identifying the repeated patterns. Therefore, the compression effect does not depend on the quality of the statistical model, nor is it restricted by the entropy of a source. It can, therefore, often achieve a better compression ratio than the methods based on a statistical model.

However, there are other issues to be considered. For example, how would certain string patterns be identified? What are the good techniques that can be used to check whether a symbol is in the dictionary? Different choices of patterns can lead to different compression results. What should we do if the dictionary expands in size too quickly? Certain data structures may affect the efficiency of certain operations directly. For example, the bigger the dictionary, the longer it takes to check whether a word is in the dictionary. Some dedicated data structures are very useful, such as circular queues, heaps, hash tables, quadtrees and tries. Centred by the three representative algorithms, many algorithms have been redeveloped to achieve or improve individual aspects of them.

Dictionary algorithms have many applications and have been used in a number of commercial software programs. For example, in UNIX or Linux, commands `compress`, `uncompress`, `gzip` and `gunzip` have all used the dictionary compression methods at some stage. Since our interest lies in the approaches of dictionary algorithms, we shall look at three most popular algorithms in fundamental form, namely LZ77, LZ78 and LZW.

The algorithms are named after the authors Abraham Lempel and Jakob Ziv who published the papers in 1977 and 1978. A popular variant of LZ78, known as (basic) LZW, was published by Terry Welch in 1984. There are numerous variants of LZ77 and LZ78/LZW. We focus on and discuss the ideas of each of these algorithms here.

7.1 Patterns in a string

The main part of a dictionary compression algorithm is to identify repetition pattern from a string. To understand the issues, we first review some prelimi-

[1]Note that there can be static dictionaries.

naries for string matching methods.

We first look at a matching problem using an example below.

Example 7.1 *Given a string* `ABBBAABABA`*, find the longest repeated pattern.*

Before solving this problem, we first need to make it more specific. For example, what is a repeated pattern? How do we define the length of a pattern?

By repeated pattern, we mean a substring of the given string that occurs at least twice in the given string. Since a pattern is a substring, we define the length of the pattern as the length of the substring.

For example, substring `BA` is a repeated pattern because it occurs three times in the given string and is of length 2 since it consists of two symbols. We highlight the recurrence of `BA`: `ABBBAABABA`.

There may be other repeated patterns in a string such as `AB`: `ABBBAABABA`.

If a pattern is given as well as the string, the length of the pattern is fixed. However, when the pattern is not specified, finding the longest pattern can be computationally expensive. It may require searching patterns of all possible lengths, e.g. the repeated pattern of length 1, of length 2, and so on.

Fortunately, the repeated patterns in dictionary compression algorithms are defined as a `word in the dictionary`. The longest patterns merely mean the longest words in the dictionary. If it is not a word, the current string will be defined as a new word, i.e. a new entry to be added to the dictionary. Therefore, the problem is easier than the *all repeated pattern* problem.

We still need to find the longest pattern in a string though. One easy method to solve this problem is to maintain a variable for the substring seen so far. Let the variable be *word*, and the newly input symbol be x. Now *word* is a recurrent pattern if it is in the dictionary, and *word* is the longest pattern seen so far if $word + x$ is not in the dictionary.[2] At the same time, $word + x$ can be inserted into the dictionary as a new entry.

Of course, $word + x$ is not the longest pattern if $word + x$ is in the dictionary. The search for the longest pattern should be continued by inputting the next symbol.

For example, suppose *word* contains `AB`. The next symbol input x is `B`. Suppose `AB` is in the dictionary but `ABB` is not. We know then that `AB` is the longest string in the dictionary.

7.2 LZW coding

People often find that LZW algorithms are easier to understand and are the most popular ones. We therefore study LZW coding first.

7.2.1 Encoding

In theory, the dictionary is built from scratch and is empty initially. However, if we know the alphabet of a source, the alphabet and other commonly used

[2]The operator '+' means concatenation for string operants.

symbols are stored as the first 256 entries in the dictionary. In other words, the dictionary usually contains 256 entries (e.g. ASCII codes) of single characters initially.

The main idea of the LZW encoding is to identify a longest pattern for each accumulated segment of the source text and encode them by the indices in the dictionary. If no match is found in the dictionary, the segment will become a new entry to the dictionary. There will be a match found in the dictionary if the same segment is seen next time.

The encoding algorithm is:[3]

Algorithm 7.1 LZW encoding

```
word ← ‘’
while not EOF do
   x ← read_next_character()
   if word + x is in the dictionary then
      word ← word + x
   else
      output the dictionary index for word
      add word + x to the dictionary
      word ← x
   end if
end while
output the dictionary index for word
```

The following 2 examples show how the algorithm works.

Example 7.2 *Trace the operations of the LZW algorithm on the input string ACBBAAC.*

Suppose the first 256 places in the dictionary have been filled initially with symbols as below:

```
Dictionary:
        1 2 3 4 5 6 7 8 9 10 11 12 13 14 15 16 17 18 19 20
        A B C D E F G H I J  K  L  M  N  O  P  Q  R  S  T

21 22 23 24 25 26 27      ...   256
U  V  W  X  Y  Z  space   ...   9
```

Solution

1. Initial step:

```
word: ‘’
```

[3]Note: the addition sign + in the algorithm means *concatenating*, e.g. $word + x$ means appending the character x to the string in *word*.

2. Symbols to be read: ACBBAAC

```
read next_character x: A (in dictionary)
word+x: A
word  : A
```

3. Symbols to be read: CBBAAC

```
read next_character x: C
word+x: AC (not in dictionary)
output: 1
new entry of the dictionary:  257
                              AC
word  : C
```

4. Symbols to be read: BBAAC

```
read next_character x: B
word+x: CB (not in dictionary)
output: 3
new entry of the dictionary:  258
                              CB
word  : B
```

5. Symbols to be read: BAAC

```
read next_character x: B
word+x: BB (not in the dictionary)
output: 2
new entry of the dictionary:  259
                              BB
word  : B
```

6. Symbols to be read: AAC

```
read next_character x: A
word+x: BA (not in the dictionary)
output: 2
new entry of the dictionary:  260
                              BA
word  : A
```

7. Symbols to be read: AC

```
read next_character x: A
word+x: AA (not in the dictionary)
```

```
output: 1
new entry of the dictionary:  261
                              AA
word  : A
```

8. Symbols to be read: C

```
read next_character x: C
word+x: AC (in dictionary)
word  : AC
```

9. Symbols to be read: none

```
read next_character x: EOF (i.e. end of file)
output: 257
```

So the total output (the compressed file) is: 1 3 2 2 1 257.

The new entries of the dictionary are:

```
257  258  259  260  261
AC   CB   BB   BA   AA
```

Example 7.3 *Encode* `AAABAABBBB` *by tracing the LZW algorithm.*

Solution Suppose the first 256 places in the dictionary have been filled initially with symbols as below:

```
Dictionary:
        1 2 3 4 5 6 7 8 9 10 11 12 13 14 15 16 17 18 19 20
        A B C D E F G H I J  K  L  M  N  O  P  Q  R  S  T

21 22 23 24 25 26 27      ...    256
U  V  W  X  Y  Z  space   ...    9
```

1. Initial step:

```
word: ‘’
```

We show the actions taken and the variable values updated by the encoding algorithm on completion of each iteration of step 2, as follows:

2. Symbols to be read: AAABAABBBB

```
Input next_character (x): A
word+x: A
word  : A
```

3. Symbols to be read: AABAABBBB

```
Input next_character (x): A
word+x: AA
Output: 1
Dictionary (new entries): 257
                          AA
word  : A
```

4. Symbols to be read: ABAABBBB

```
Input next_character (x): A
word+x: AA
word  : AA
```

5. Symbols to be read: BAABBBB

```
Input next_character (x): B
word+x: AAB
Output: 257
Dictionary (new entries): 257 258
                          AA  AAB
word  : B
```

6. Symbols to be read: AABBBB

```
Input next_character (x): A
word+x: BA
Output: 2
Dictionary (new entries): 257 258 259
                          AA  AAB BA
word  : A
```

7. Symbols to be read: ABBBB

```
Input next_character (x): A
word+x: AA
word  : AA
```

8. Symbols to be read: BBBB

```
Input next_character (x): B
word+x: AAB
word  : AAB
```

9. Symbols to be read: BBB

```
Input next_character (x): B
word+x: AABB
Output: 258
Dictionary (new entries): 257 258 259 260
                          AA  AAB BA  AABB
word  : B
```

10. Symbols to be read: BB

```
Input next_character (x): B
word+x: BB
Output: 2
Dictionary (new entries): 257 258 259 260  261
                          AA  AAA BA  AABB BB
word  : B
```

11. Symbols to be read: B

```
Input next_character (x): B
word+x: BB
word  : BB
```

12. Symbols to be read: (none)

```
Input next_character (x): EOF (i.e. end of file)
/* goes to step 3. */
Output: 261
```

So the compressed output for AAABAABBBB is

```
1 257 2 258 2 261
```

The new entries of the dictionary are:

```
257 258 259 260  261
AA  AAB BA  AABB BB
```

7.2.2 Decoding

Similarly, the decoder builds its own dictionary as it reads the tokens, i.e. the encoded data, one by one from the compressed file.

The algorithm now is:

Algorithm 7.2 LZW decoding

```
read a token x from the compressed file
look up dictionary for element at x
output element
word ← element
while not EOF do
    read x
    look up dictionary for element at x
    if there is no entry yet for index x then
        element ← word + firstCharOfWord
    end if
    output element
    add word + firstCharOfElement to the dictionary
    word ← element
end while
```

Example 7.4 *Show the operations of the decompress algorithm on the input tokens 1 3 2 2 1 257, which is the compressed result in Example 7.2.*

Solution

1. (Initial step) Tokens to be read: 1 3 2 2 1 257

```
read token x: 1
element: A
output: A
word : A
```

2. Tokens to be read: 3 2 2 1 257

```
read token x: 3
element: C
output: C
add new entry of dictionary: 257
                             AC
word : C
```

3. Tokens to be read: 2 2 1 257

```
read token x: 2
element: B
output: B
add new entry of dictionary: 258
                             CB
word : B
```

4. Tokens to be read: 2 1 257

```
read token x: 2
element: B
output: B
add new entry of dictionary: 259
                             BB
word : B
```

5. Tokens to be read: 1 257

```
read token x: 1
element: A
output A
add new entry of dictionary: 260
                             BA
word : A
```

6. Tokens to be read: 257

```
read token x: 257
element: AC
output AC
add new entry of dictionary: 261
                             AA
word : AC
```

7. Token to be read: (none)

```
read token x: EOF (i.e. end of compressed file)
(end)
```

So the total output is ACBBAAC.
The new entries of the dictionary are:

257	258	259	260	261
AC	CB	BB	BA	AA

Example 7.5 *Decode 1 257 2 258 2 261 obtained from Example 7.3 and trace the activities of the decoding algorithm.*

Suppose again the first 256 places in the dictionary have been filled initially with symbols as below:

```
Dictionary:
        1 2 3 4 5 6 7 8 9 10 11 12 13 14 15 16 17 18 19 20
        A B C D E F G H I J  K  L  M  N  O  P  Q  R  S  T

21 22 23 24 25 26 27      ...    256
U  V  W  X  Y  Z  space  ...    9
```

Solution

1. Tokens to be read: 1 257 2 258 2 261

   Initial steps:

   ```
   Read next_token (x): 1
   output: A
   word: A
   ```

2. Tokens to be read: 257 2 258 2 261

   (Iterations begin):

   ```
   Read next_token (x): 257
   Look up the dictionary and find there is no entry yet for 257
   element: AA /* element=word+first_char_of_word */
   output element: AA
   Dictionary (new entries so far): 257
                                    AA
   word: AA
   ```

3. Tokens to be read: 2 258 2 261

   ```
   Read next_token (x): 2
   Look up the dictionary and output element: B
   Dictionary (new entries so far): 257 258
                                    AA  AAB
   word: B
   ```

4. Tokens to be read: 258 2 261

```
Read next_token (x): 258
Look up the dictionary and output element: AAB
Dictionary (new entries so far): 257 258 259
                                  AA  AAB BA
word: AAB
```

5. Tokens to be read: 2 261

```
Read next_token (x): 2
Look up the dictionary and output element: B
Dictionary (new entries so far): 257 258 259 260
                                  AA  AAB BA  AABB
word: B
```

6. Tokens to be read: 261

```
Read next_token (x): 261
Look up the dictionary and find no entry for 261
element: BB /* element=word+first_char_of_word */
output element: BB
Dictionary (new entries so far): 257 258 259 260  261
                                  AA  AAB BA  AABB BB
word: BB
```

7. Tokens to be read: (none)

```
Read next_token (x): EOF (i.e. end of compressed file)
(end)
```

So the decoded message is AAABAABBBB.
Dictionary (new entries) is:

```
257 258 259 260  261
AA  AAB BA  AABB BB
```

Observation

1. The compression algorithm and the decompression algorithm build an identical dictionary independently. The advantage of this is that the compression algorithm does not have to pass the dictionary to the decompressor.
2. The size of the dictionary may grow so quickly that an effective method of maintaining the dictionary would be essential.

7.3 LZ77 family

In this approach, the dictionary to use is a *portion* of the previously seen input file. The proportion of the input string is decided by an imagined sliding window which can be shifted from left to right. The window is maintained to define dynamically the dictionary part and to scan the input sequence of symbols. As the window sliding from the left to right, the content of the dictionary and the portion of the input text in which patterns are sought are updated. This imitates the situation when a compressor scans the source text segment by segment.

The window is divided into two consecutive halves, the first half is called the *History buffer* (H for short) or the *Search buffer* which contains a portion of the recently seen symbol sequence. The second half of the window is called the *Lookahead buffer* (L for short) which contains the next portion of the sequence to be encoded. The term *buffer* is used to mean a storage for some temporary data.

The size of each buffer is usually fixed in advance. In practical implementation, the History buffer is some thousands of bytes long and the Lookahead buffer is only tens of bytes long.

Let l_H and l_L be the size of H and L respectively.

Figure 7.1 shows an example, where H = 'ry␣based␣compres', a History buffer of size $l_H = 16$ bytes (1 byte for each symbol), and L = 'sion␣compress', a Lookahead buffer of size $l_L = 12$. The sequence of symbols in (and before) H has been seen and compressed but the sequence in (and after) L is to be compressed. You can imagine that the window moves from left to right (or the entire text sequence moves from right to left) from time to time during the compression process.

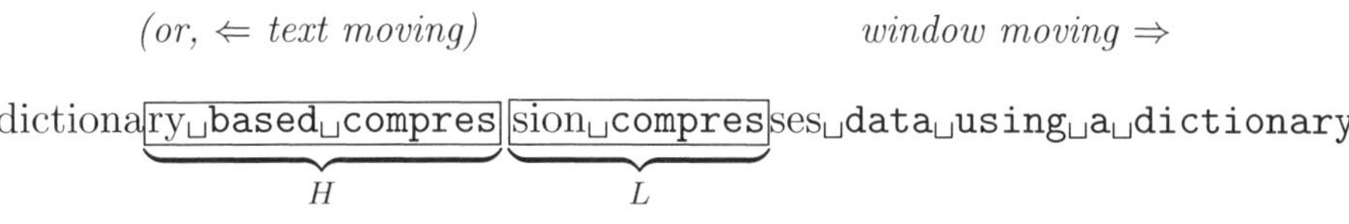

Figure 7.1: History buffer and Lookahead buffer

Similar to the LZW algorithm, the input of the compression algorithm is a sequence of the source symbols and the output of the compression is a triple codeword $\langle f, l, c\rangle$ called *token*. If the tokens are regarded as of fixed length, the LZ77 compression algorithms belong to the *variable-to-fixed* coding scheme.

7.3.1 Prefix match

Before moving on to discuss the LZ77 algorithm, let us first clarify a few concepts related to the codeword $\langle f, l, c\rangle$.

An array of n symbols is called a *string* of length n. A substring is a number ($\leq n$) of consecutive symbols in the string. A prefix of a string is a substring of any length beginning with the first symbol of the string.

Example 7.6 *Consider string* `BABC`. *Its substrings are* `B, A, B, C, BA, AB, BC, BAB, ABC, BABC`. *The possible prefixes are* `B, BA, BC, BAB, BABC`.

Now consider two strings: H = '`ABBBAABABA`' and L = '`BABC`'. A prefix of L may occur in the string H, i.e. being identical to a substring of H. Such an occurrence is called a prefix *match* (of L) in H by our definition. For instance, `B, BA` and `BAB` are three prefix matches found in H.

We are often more interested in the longest prefix match, such as the `BAB` in H in the example. For convenience, we use the term *prefix match* to mean the longest substring (or substrings) in the History buffer H that matches a prefix in L. We highlight the prefix match in H and L below. Index 7 is also highlighted since it indicates the match location in H.

1	2	3	4	5	6	7	8	9	10	1	2	3	4
A	B	B	B	A	A	B	A	B	A	B	A	B	C

The longest prefix length is called the *length* of the match (*mlength* for short) which is 3 in this example. The start position of the match is the index 7 in H. We indicate this position by the so-called *offset* which is the distance from the right edge of H. The *offset* is 4 in the example. The first mismatching symbol is `C` following the match in L (see the diagram below).

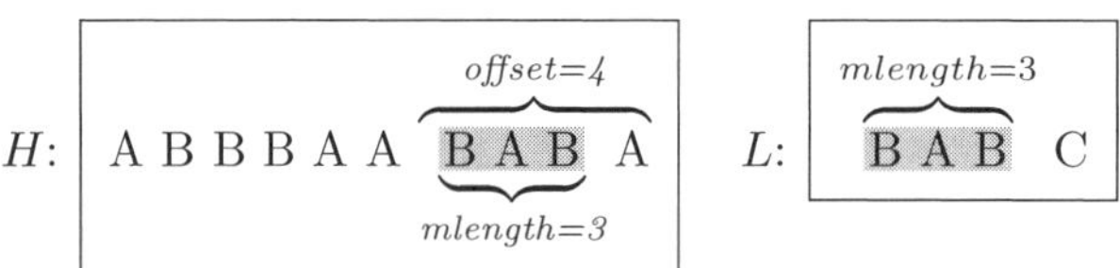

These three values form the token $\langle f, l, c\rangle$ in the LZ77 algorithm below, where f is the offset, i.e. the location from the right edge of the H; l is the length of the match found in H and c is the *immediate* mismatching symbol after the prefix in the L.

We shall need the concept of such a *prefix match*, the *length* of the match and *offset* in understanding the LZ77 algorithm below.

7.3.2 A typical compression step

Suppose that H, the History buffer, has l_H bytes (characters) that have been seen and encoded; L, the Lookahead buffer, has at most l_L characters, which have been seen but not yet encoded. There are many algorithms in the LZ77 family but a typical encoding algorithm can be outlined below:

1. Read from the input characters until L is full.
2. Scan H from right to left searching H for a prefix match (as described in Section 7.3.1).
   If more than one match is found in the H, take the right most prefix match, i.e. the longest and the first one from the right in H.[4]

[4] i.e. the one located furthest from the left in H.

window moving ⇒

dictionary␣b`ased␣compression`|`␣compresses␣`data␣using␣a␣dictionary

H *L*

Figure 7.2: A match is found with an offset length 12 and match length 9

3. If a match of length $l \geq 2$ characters is found (see Figure 7.2) with an offset of f bytes, then output token $\langle f, l, c\rangle$, where c is the first mismatching symbol following the match.

   Slide the window (H and L together) f characters to the right (i.e. shift to the left the first l characters out of H, and the first l characters in L into H.)

4. If no match is found then output $\langle 0, 0, \text{ASCII(c)}\rangle$.

   Slide the window one character to the right.

In some cases, the last element of the triple, c, may be unnecessary and can be saved.

Example 7.7 *Starting with a H of size 16 bytes which contains the characters* `ased␣compression` *(where* ␣ *represents a space) and an empty Lookahead buffer, show the output and the final state of the History buffer if the following characters are next in the input, for a Lookahead buffer of size 12:* `␣compresses␣data␣using␣a␣dictionary`.

Solution We refer to Figure 7.2 and trace what happens step by step following the LZ77 encoding algorithm.

1. Load the input string into L, H (left) and L (right) now contain the following:

   `ased␣compression`|`␣compresses␣`data␣using␣a␣dictionary

   A prefix match '`␣compress`' is found and we record the offset $f = 12$ (bytes) and the match length $l = 9$ (bytes).

   Output: $\langle 12, 9\rangle$.

   Slide the window nine characters to the right. H (left) and L (right) now contain:

   `ression␣compress`|`es␣data␣using␣a␣dictionary`

2. A match 'es' is found,[5] so we record the offset 3 and the match length 2.

   Output: $\langle 3, 2\rangle$.

   Slide the window two characters to the right. H (left) and L (right) now contain:

   `ssion␣compresses`|`␣data␣using␣`a␣dictionary

[5]In fact, 'es' is found in two places but we only consider the first (right most) match here.

3. Only a one-symbol match '␣' is found so:

   Output: the pointer $\langle 0,$ `ASCII('␣')`$\rangle$.

   Slide the window one character to the right. H (left) and L (right) now contain:

   `|sion␣compresses␣|data␣using␣a|␣dictionary`

4. No match is found so:

   Output: the pointer $\langle 0,$ `ASCII('d')`$\rangle$.

   Slide the window one character to the right. H (left) and L (right) now contain:

   `|ion␣compresses␣d|ata␣using␣a␣|dictionary`

5. No match is found so:

   Output: the pointer $\langle 0,$ `ASCII('a')`$\rangle$.

   Slide the window one character to the right. H (left) and L (right) now contain:

   `|on␣compresses␣da|ta␣using␣a␣d|ictionary`

6. No match is found so:

   Output: the pointer $\langle 0,$ `ASCII('t')`$\rangle$.

   Slide the window one character to the right. H (left) and L (right) now contain:

   `|n␣compresses␣dat|a␣using␣a␣di|ctionary`

7. Only one symbol match `'a'` is found so:

   Output: the pointer $\langle 0,$ `ASCII('a')`$\rangle$.

   Slide the window one character to the right. H (left) and L (right) now contain:

   `|␣compresses␣data|␣using␣a␣dic|tionary`

   ⋮

   and so on.

To write the algorithm, we need to define certain variables to find the offset, the match and shift of the contents dynamically. This can be summarised in the following diagram:

Suppose the entire text of N bytes long is stored in an array $S[1 \cdots N]$, the size of the sliding window is W. We define the following variables for implementation convenience:

- $S[m]$ is the first character in the match found in H
- p is the first character in L

- The offset $f = p - m$
- l is the match length
- The first mismatching character is $S[p + l]$

This can be seen clearly in Figure 7.3.

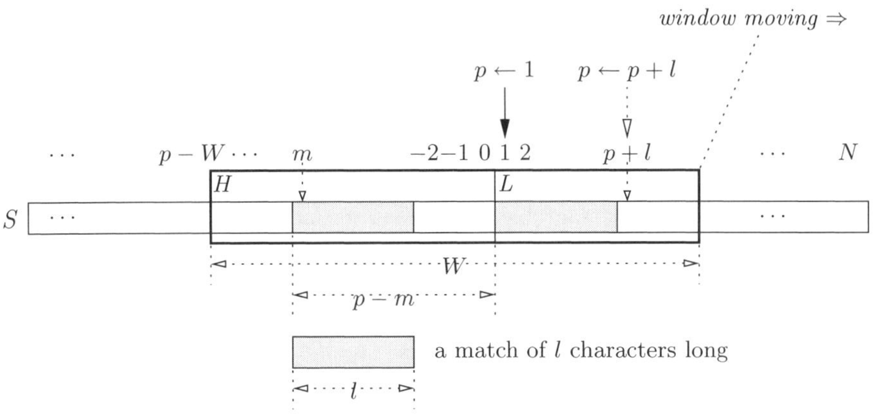

$L = S[p \cdots l_L]$ $S[m]$ is the first character in the match

Figure 7.3: $S[1 \cdots N]$ for LZ77 encoding

We now can write the Algorithm 7.3.

Algorithm 7.3 LZ77 encoding

INPUT: symbol sequence, window size W
OUTPUT: token sequence in $\langle f, l, c \rangle$ form

```
set p ← 1
while not EOF do
    find the longest match of l bytes for S[p···l] in H = S[p − W···(p − 1)]
    and the first matching character is S[m]
    output the triple token ⟨p − m, l, S[p + l]⟩
    p ← p + l + 1 {prepare for the next shift}
end while
```

Observation

1. The History buffer is initialised to have no characters in it. Hence the first few characters are coded as raw ASCII codes (and the overhead) as the History buffer fills up. This may result in expansion of the compressed file rather than compression at the beginning.
2. In common with the adaptive Huffman algorithm, LZ77 is an adaptive algorithm and may start with an empty model.

3. The LZ77 encoding algorithm can only identify the patterns in the source file which are close enough to each other, because it can only compare the Lookahead buffer to the recently seen text in the History buffer, but not to the text that was input a long time ago.

7.3.3 Decompression algorithm

The decoder also maintains a buffer of the same size as the encoder's window. However, it is much simpler than the encoder because there is no matching problems to be dealt with. The decoder reads a token and decides whether the token represents a match.

If it is a match, the decoder will use the offset and the match length in the token to reconstruct the match. Otherwise, it outputs the symbol in ASCII code in the token.

We use an example to show how this works.

Example 7.8 *Suppose the buffer contains the following decoded symbols:* `Dictionary␣based␣compression`.

The decoder reads the following tokens one at each iteration: $\langle 12, 9\rangle\langle 3, 2\rangle$ $\langle 0, ASCII(\text{'␣'})\rangle\langle 0, ASCII(\text{'d'})\rangle\langle 0, ASCII(\text{'a'})\rangle$ $\langle 0, ASCII(\text{'t'})\rangle\langle 0, ASCII(\text{'a'})\rangle$

Solution

1. Read the next token $\langle 12, 9\rangle$, which is a match. The decoder uses the offset (the first number in the token f) to find the first symbol of the match and uses the match length (the second number in the token l) to decide the number of symbols to copy.

   So we have the following process:

   (a) Count 12 from the end of the buffer and find the symbol '␣'.

   (b) Copy the next nine symbols one by one (see the shade characters) and update the buffer:

   ```
   Dictionary␣based␣compression␣
   Dictionary␣based␣compression␣c
   Dictionary␣based␣compression␣co
   Dictionary␣based␣compression␣com
   Dictionary␣based␣compression␣comp
   Dictionary␣based␣compression␣compr
   Dictionary␣based␣compression␣compre
   Dictionary␣based␣compression␣compres
   Dictionary␣based␣compression␣compress
   ```

2. Read $\langle 3, 2\rangle$, find the position with offset 3, copy the next 2 characters, and the buffer becomes:

   `based␣compression␣compresses`

3. Read $\langle 0, \texttt{ASCII('␣')}\rangle$, which is not a match so output: '␣'.
   The buffer becomes:
   `sed␣compression␣compresses␣`
4. Read $\langle 0, \texttt{ASCII('d')}\rangle$, which is not a match so output: 'd' (i.e. add 'd' into the buffer).
   The buffer becomes:
   `ed␣compression␣compresses␣d`
5. Read $\langle 0, \texttt{ASCII('a')}\rangle$, which is not a match so output: 'a'.
   The buffer becomes:
   `d␣compression␣compresses␣da`
6. Read $\langle 0, \texttt{ASCII('t')}\rangle$, which is not a match so output: 't'.
   The buffer becomes:
   `␣compression␣compresses␣dat`
7. Read $\langle 0, \texttt{ASCII('a')}\rangle$, which is not a match so output: 'a'.
   The buffer becomes:
   `compression␣compresses␣data`

   $\vdots$

   and so on.

Algorithm 7.4 summarises the decoding process.

Algorithm 7.4 LZ77 decoding

INPUT: token sequence in $\langle f, l, c\rangle$ form, window size W
OUTPUT: symbol sequence

set $p \leftarrow 1$
while not EOF **do**
  read next token $\langle f, l, c\rangle$
  set $S[p \cdots (p+l-1)] \leftarrow S[(p-f) \cdots (p-f+l-1)]$
  $S[p+l] \leftarrow c$
  $p \leftarrow p+l+1$ {prepare for the next shift}
end while

Observation

1. The decompression algorithm builds up the same History buffer as the compression algorithm, and decodes tokens with reference to the History buffer. The decompression algorithm decides whether the next character is a 'real' index or a raw symbol, this depends on the first component of the token. If the first is a 0, the next character is a raw symbol.

2. The algorithm is *asymmetric* since compression is slower than decompression. The compression algorithm involves searching for a match, which is computationally intensive, but decompression only involves reading out values from the History buffer.

7.3.4 Implementation issues

Size of the two buffers

In LZ77, there is an important design decision to be made concerning the values of l_H and l_L:

1. Choosing a large History buffer means it is likely that matches are found, but the offsets will be larger. A smaller buffer means smaller pointers but less chance of finding a match.
2. Choosing a small Lookahead buffer means a quick search for a prefix, but the chance of the match found will be limited.

The basic LZ77 method has been improved in many ways since the 1980s. For example, a variable-sized offset and length component, i.e. f and l in the token $\langle f, l, c \rangle$, can be used to improve the compression performance.

Another improvement is to increasing the size of buffers. As we know, increasing the size of the History buffer can have better chances of finding matches. The more matches found, the better compression ratio can be achieved. However, a bigger History buffer may in general slow down the search. Useful work, therefore, has been focusing on use of more efficient data structures such as tries and hashing techniques.

Circular queues

A circular queue is a dynamic queue structure implemented by an array which is a static structure. The front and rear of the queue are pointed by the two variables storing the indices of the position. This eases the operation of appending elements at the rear and removing elements at the front of the queue.

Given a sequence of symbols in the source string, adding a few elements becomes the equivalent of updating the index value of a front variable.

7.4 LZ78 family

A big problem of LZ77 is that it cannot recognise the patterns occurring some time ago because they may have been shifted out from the History buffer. In this situation, the patterns are ignored and no matches are found. This leads to expansion instead of compression of this part of the source text by outputting a token triple for a single character.

To extend the 'memory' of the patterns, LZ78 algorithms are developed to maintain a dictionary that allow patterns to remain as entries permanently during the whole encoding process.

LZ78 also requires one component less in triple tokens compared to that in LZ77 and only outputs pair tokens instead. A pair token is defined as $\langle f, c\rangle$ where f represents the offset which indicates the starting position of a match, and c is the character of the next symbol to the match in the source text. The length of the match is included in the dictionary, so there is no need to include the information in the code.

Typical LZ78 compression algorithms use a *trie*[6] to keep track of *all* the patterns seen so far. The dictionary D contains a set of pattern entries, which are indexed from 0 onwards using integers. Similar to LZ77, the index corresponding to a word in the dictionary is called the token. The output of the encoding algorithm is a sequence of tokens only. If a symbol is not found in the dictionary, the token $\langle 0, x\rangle$ will be output which indicates a concatenation of the null string and x. Initially the dictionary `D` is normally loaded with all 256 single character strings. Each single character is represented simply by its ASCII code. All subsequent entries are given token numbers 256 or more.

7.4.1 Encoding

Let *word* be the currently matched string. Initially *word* is empty.

The encoding algorithm is as follows.

Algorithm 7.5 LZ78 encoding

INPUT: string of symbols, dictionary with an empty entry at index 0
OUTPUT: sequence of tokens $\langle index(word),\ c\rangle$, updated dictionary

```
while not EOF do
   word ← empty
   c ← next_char()
   while word + c is in the Dictionary do
      word ← word + c
      c ← next_char()
   end while
   output token ⟨index(word), c⟩
   {where index(word) is the index of word in the dictionary}
   add word + c into the dictionary at the next available location
end while
```

Example 7.9 *Show step by step the encoding operation of LZ78 on the input string:*

`a␣date␣at␣a␣date`

Solution We trace the values of *word* (w), c, output and the Dictionary (D) for each iteration (i). Initially, the dictionary is empty.

[6] This is a commonly used data structure for strings and is similar to trees.

										D					
i	w	c	w+c	output	0	1	2	3	4	5	6	7	8	9	10
1		a	a	0, a		a									
2		␣	␣	0, ␣		a	␣								
3		d	d	0, d		a	␣	d							
4		a	a												
	a	t	at	1, t		a	␣	d	at						
5		e	e	0, e		a	␣	d	at	e					
6		␣	␣												
	␣	a	␣a	2, a		a	␣	d	at	e	␣a				
7		t	t	0, t		a	␣	d	at	e	␣a	t			
8		␣	␣												
	␣	a	␣a												
	␣a	␣	␣a␣	6, ␣		a	␣	d	at	e	␣a	t	␣a␣		
9		d	d												
	d	a	da	3, a		a	␣	d	at	e	␣a	t	␣a␣	da	
10		t	t												
	t	e	te	7, e		a	␣	d	at	e	␣a	t	␣a␣	da	te

7.4.2 Decoding

The decoding algorithm reads an element of the tokens at a time from the compressed file and maintains the dictionary in a similar way as the encoder.

Let $\langle x, c\rangle$ be a compressed token pair, where x is the next codeword and c the character after it.

Algorithm 7.6 LZ78 decoding

INPUT: sequence of tokens in $\langle x,\ c\rangle$ format,
dictionary with an empty entry at index 0

OUTPUT: string of decoded symbols, updated dictionary

```
while not EOF do
    x ← next_codeword()
    c ← next_char()
    output dictionary_word(x) + c
    add dictionary_word(x) + c into dictionary at the next available location
end while
```

Example 7.10 *Show step by step the decoding operation of LZ78 on the input tokens:* 0a 0␣ 0d 1t 0e 2a 0t 6␣ 3a 7e

Solution We trace the values of x, c, output and the dictionary (D) for each iteration (i). Initially, the dictionary is empty. In the table below, we use *dictionary_word*(x) to mean the word at index x in the dictionary ($w(x)$ for short).

					D										
i	x	c	$w(x)+c$	output	0	1	2	3	4	5	6	7	8	9	10
1	0	a	a	a		a									
2	0	␣	␣	␣		a	␣								
3	0	d	d	d		a	␣	d							
4	1	t	at	at		a	␣	d	at						
5	0	e	e	e		a	␣	d	at	e					
6	2	a	␣a	␣a		a	␣	d	at	e	␣a				
7	0	t	t	t		a	␣	d	at	e	␣a	t			
8	6	␣	␣a␣	␣a␣		a	␣	d	at	e	␣a	t	␣a␣		
9	3	a	da	da		a	␣	d	at	e	␣a	t	␣a␣	da	
10	7	e	te	te		a	␣	d	at	e	␣a	t	␣a␣	da	te

We now have the original string `a␣date␣at␣a␣date` back.

Observation

1. LZ78 has made some improvement over LZ77. For example, in theory the dictionary can keep the patterns forever after they have been seen once. In practice, however, the size of the dictionary cannot grow indefinitely. Some patterns may need to be reinstalled when the dictionary is full.
2. The output codewords contain one less component than those in LZ77. This improves the data efficiency.
3. LZ78 has many variants and LZW is the most popular variance to LZ78, where the dictionary begins with all the 256 initial symbols and the output pair is simplified to output only a single element.

7.5 Applications

One of the Unix utilities `compress` is a widely used LZW variant.

- The number of bits used for representing tokens is increased gradually as needed. For example, when the token number reaches 255, all the tokens are coded using 9 bits, until the token number 511 ($2^9 - 1$) is reached. After that, 10 bits are used to encode tokens and so on.
- When the dictionary is full (i.e. the token number reaches its limit), the algorithm stops adapting and only uses existing words in the dictionary. At this time, the compression performance is monitored, and the dictionary is *rebuilt* from scratch if the performance deteriorates significantly.
- The dictionary is represented using a *trie* data structure.

GIF (Graphics Interchange Format) is a lossless image compression format introduced by CompuServe in 1987.

- Each pixel of the images is an index into a table that specifies a colour map.

- The colour table is allowed to be specified along with each image (or with a group of images sharing the map).
- The table forms an uncompressed prefix to the image file, and may specify up to 256 colour table entries each of 24 bits.
- The image is really a sequence of 256 different symbols and is compressed using the LZW algorithm.

V.42bis is an ITU-T standard commmunication protocol for telephone-line modems that applies the LZW compression method.

- Each modem has a pair of dictionaries, one for incoming data and one for outgoing data.
- The maximum dictionary size is often negotiated between the sending and receiving modem as the connection is made. The minimum size is 512 tokens with a maximum of six characters per token.
- Those tokens to be used infrequently may be deleted from the dictionary.
- The modem may switch to transmitting uncompressed data if it detects that compression is not happening (e.g. if the file to transmit has already been compressed).
- The modem may also request the called modem to discard the dictionary, when a new file is to be transmitted.

7.6 Comparison

We have studied all the important lossless compression algorithms for compressing text. Some of them are static and others are adaptive. Some use fixed length codes, others use variable length codes. Some compression algorithms and decompression algorithms use the same model, others use different ones. We summarise these characteristics below:

Algorithm	Adaptive	Symmetric	Type
Run-length	n	y	variable to fixed
Huffman	n	y	fixed to variable
Adaptive Huffman	y	y	fixed to variable
Arithmetic	y	y	variable to variable
LZ77	y	n	variable to fixed
LZW	y	y	variable to fixed

Summary

Dictionary compression algorithms use no statistical models. They focus on the memory on the strings already seen. The memory may be an explicit dictionary that can be extended infinitely, or an implicit limited dictionary as sliding windows. Each seen string is stored into a dictionary with an index. The indices of all the seen strings are used as codewords. The compression and decompression algorithm maintains individually its own dictionary but the two dictionaries are identical. Many variations are based on three representative families, namely LZ77, LZ78 and LZW. Implementation issues include the choice of the size of the buffers, the dictionary and indices.

Learning outcomes

On completion of your studies in this chapter, you should be able to:

- explain the main ideas of dictionary-based compression
- describe compression and decompression algorithms such as LZW, LZ77 and LZ78
- list and comment on the main implementation issues for dictionary-based compression algorithms.

Exercises

E7.1 Using string `abbaacabbabbb#` (where `#` represents the end of the string) as an example, trace the values of word, x and the dictionary in running the basic LZW *encoding* and *decoding* algorithms, where word is the accumulated string and x is the character read on each iteration.

E7.2 Suppose the input for encoding is a string `aabbaaaccdee`. Demonstrate how the simplified version algorithms of LZ77, LZ78 and LZW work step by step.

E7.3 Suppose the History buffer is seven characters long and the Lookahead buffer is four characters long. Illustrate how LZ77 compression and decompression algorithms work by analysing the buffer content on each iteration.

E7.4 For LZ78 and LZW, show what the dictionaries look like at the completion of the input sequence in the following format.

Dictionary address (in decimal)	Dictionary entry
0	?, ?
1	?, ?
2	?, ?
3	?, ?
⋮	

E7.5 Describe a simplified version of the LZ77, LZ78 and LZW algorithms. Analyse the type of the simplified algorithms in terms of, for example, whether they are *static or adaptive* and *fixed-to-variable*, *variable-to-fixed* or *variable-to-variable*.

E7.6 Analyse the so-called LZSS algorithm below and discuss the advantages and disadvantages.

Algorithm 7.7 LZSS encoding

INPUT: symbol sequence
OUTPUT: $\langle f, l \rangle$, or $\langle 0, f, l \rangle$

while not EOF **do**
  determine $\langle f, l \rangle$ corresponding to the match in L
end while
if $sizeOf(\langle f, l \rangle) \geq sizeOf(\text{match})$ **then**
  output the token $\langle 1,\ first\_character(L) \rangle$
  shift the entire buffer content by one position
else
  output $\langle 0, p, l \rangle$
  shift the entire buffer content by l positions
end if

Laboratory

L7.1 Implement a simple version of the LZW encoding and decoding algorithms.

L7.2 Implement a simple version of the LZ77 encoding and decoding algorithm.

L7.3 Investigate, experiment and comment on the performance of your programs.

L7.4 Implement Algorithm 7.8, an alternative LZ78 decoding algorithm. Discuss the difference between this and Algorithm 7.6 in terms of the performance.

Assessment

S7.1 Explain why a dictionary-based coding method such as LZ77 is said to be *adaptive* and *variable-to-fixed* in its basic form.

S7.2 One simplified version of the LZ78/LZW algorithm can be described as in Algorithm 7.9, where n is a pointer to another location in the dictionary, c is a symbol drawn from the source alphabet, and $\langle n, c \rangle$ can form a node of a linked list. Suppose that the pointer variable n also serves as the transmitted codeword, which consists of 8 bits. Suppose also the `0` address

Algorithm 7.8 LZ78-1 decoding

INPUT: sequence of tokens in $\langle x, c \rangle$ format, empty dictionary
OUTPUT: string of decoded symbols, updated dictionary

```
while not EOF do
    x ← next_codeword()
    c ← next_char()
    if x = 0 then
        output c
    else
        output dictionary_word(x) + c
    end if
    add dictionary_word(x) + c into dictionary at the next available location
end while
```

Algorithm 7.9 Another version of LZ78/LZW

INPUT:
OUTPUT:

```
n ← 0; fetch next source symbol c
if the ordered pair ⟨n, c⟩ is already in the dictionary then
    n ← dictionary address of entry ⟨n, c⟩
else
    transmit n {as a code word to decoder}
    create new dictionary entry ⟨n, c⟩ at dictionary address m
    m ← m + 1
    n ← dictionary address of entry ⟨0, c⟩
end if
return to step 1
```

entry in the dictionary is a `NULL` symbol, and the first 256 places in the dictionary are filled with single characters.

Suppose that a *binary* information source emits the sequence of symbols `110001011001011100011111` (that is `110 001 011 001 011 100 011 11` without spaces). Construct the encoding dictionary step by step and show, in the format below, the dictionary on completion of the above input sequence.

Dictionary address (in decimal)	Dictionary entry
0	?, ?
1	?, ?
2	?, ?
⋮	⋮

Bibliography

[Nel89] M. Nelson. LZW data compression. *Dr. Dobb's Journal*, 14:29–37, 1989.

[SS82a] J.A. Storer and T.G. Szymanski. Data compression via textual substitution. *Journal of the ACM*, 29:928–951, 1982.

[SS82b] J.A. Storer and T.G. Szymanski. Data Compression via Textual Substitution. *Journal of the ACM*, 29:928–951, 1982.

[Wel84] T.A. Welch. A technique for high performance data compression. *IEEE Computer*, 17(6):8–19, June 1984.

[ZL77] J. Ziv and A. Lempel. A universal algorithm for sequential data compression. *IEEE Transactions on Information Theory*, IT-23(3):337–343, May 1977.

[ZL78] J. Ziv and A. Lempel. Compression of individual sequences via variable rate coding. *IEEE Transactions on Information Theory*, IT-24(5):530–536, September 1978.

Chapter 8

Prediction and transforms

We have seen several lossless compression algorithms. In practice, these algorithms are often used together with some other methods to achieve better overall compression efficiency. This can be done by lining up several different algorithms and using them one after another. Certain changes can be made on the source distribution before applying the lossless compression algorithms. The source data can also be transformed to another domain to achieve better overall compression performance.

The preparation work before applying certain lossless compression algorithms is called *preprocessing*. Examples of typical preprocessing operations include *sampling* and *quantisation* as well as *prediction* and *transforms*, all of which depend on the source data. These techniques are mainly for *lossy* compression of audio, image and video data, although some approaches can be lossless.

In this chapter, we shall introduce mainly two common approaches of manipulating the source data in order to achieve a more favourable distribution for compression. One is the so-called *prediction* which changes the representation of the source data according to *prediction rules* based on the data already seen in the past. The other is called *transform* which changes the source data domain. We shall leave the sampling and quantisation to the next chapter, but introduce the concept of quantisation briefly.

8.1 Predictive approach

There are two issues in this approach. One is for the encoding process, the so-called *prediction rules* to map the original set of data to another for a skewed distribution. Data with a skewed distribution contain certain elements that occur much more frequently than others. The second is for the decoding process, an *inverse* formula to recover the original set of the data. It is easy to understand the predictive approach from the following example: for convenience, we use *predictive encoding* and *predictive decoding* to mean the forward process and the inverse process respectively. We may also use *encoding* and *decoding* for short respectively.

Example 8.1 *Consider the predictive approach on an array of data* $\boldsymbol{A} = (1, 1, 2, 3, 3, 4, 5, 7, 8, 8)$ *from alphabet* $(1, \cdots, 8)$.

We notice that each datum occurs a similar number of times. The frequency distribution can be found in the following table and its plot is seen relatively flat in Figure 8.1.

x (data)	1	2	3	4	5	6	7	8
f (frequency)	2	1	2	1	1	0	1	2

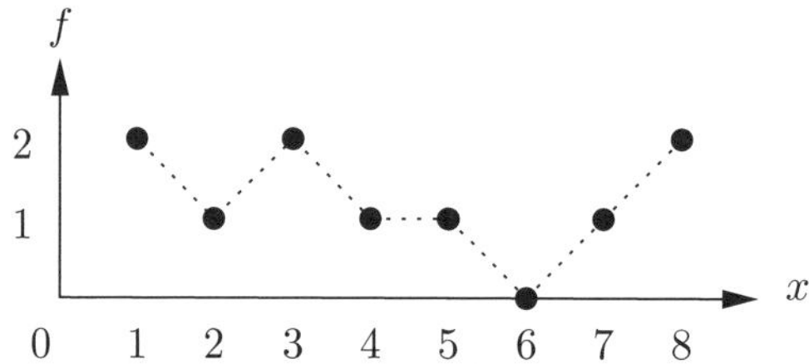

Figure 8.1: Frequency distribution of **A**

Predictive encoding

Since the frequency distribution is flat, the compression methods we have learnt so far would not be very effective to the data. Now if, for each datum from the *second position*, we add 1 to its previous datum, the sum can be used as its predicted value based on the previous item. We have the predicted values: $\mathbf{A}' = (\boxed{1}, 2, 2, 3, 4, 4, 5, 6, 8, 9)$, where the datum marked in a small box remains unchanged. Subtract **A** from **A**′ and we have for each datum, from the second position, the difference between the predicted value and the original value: $\mathbf{B} = \mathbf{A}' - \mathbf{A} = (\boxed{1}, 1, 0, 0, 1, 0, 0, -1, 0, 1)$. Obviously, a better compression may be achieved on **B** than on the original **A**, because the distribution of data in **B** is skewed compared to the original data **A**. Figure 8.2 shows the range of the data has been reduced from [1, 8] to [−1, 1], and there are more 0s now.

Figure 8.3 shows the frequency distribution before (a) and after (b) the predictive operations.

Predictive decoding

Given $\mathbf{B} = (\boxed{1}, 1, 0, 0, 1, 0, 0, -1, 0, 1)$, we can simply 'subtract the current **B** value from the previous datum in **A** and add 1' to get the original **A**. For each i, $A[i] = A[i-1] - B[i] + 1$ where $A[1] = B[1]$ initially. For example, since

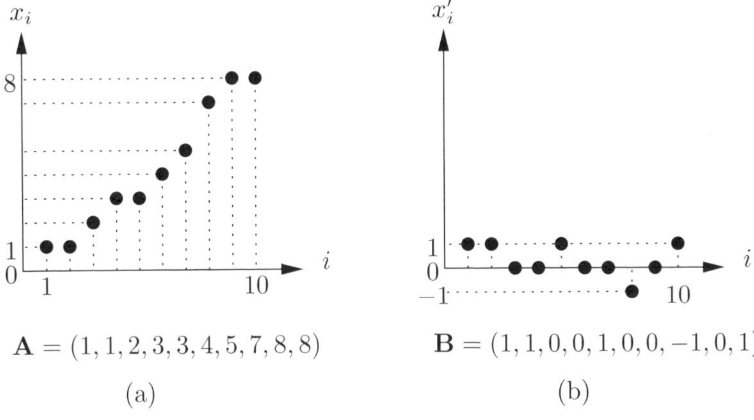

Figure 8.2: Plot of arrays **A** and **B**

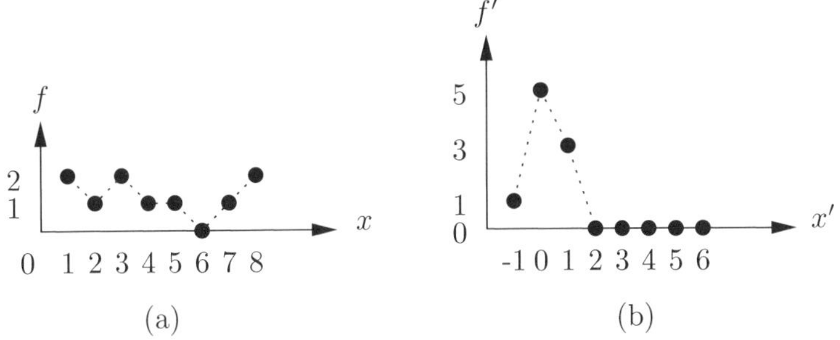

Figure 8.3: The frequency distribution for **A** (a) and **B** (b)

$B[1] = 1$, we know $A[1] = 1$. Then $A[2] = A[1] - B[2] + 1 = 1 - 1 + 1 = 1$, $A[3] = A[2] - B[3] + 1 = 1 - 0 + 1 = 2$, and so on.

As we can see from the example, the encoding process must be *reversible*. The prediction rule 'adding 1 to the previous datum' in the approach can be written as $A'[i] = A[i-1] + 1$, for $i = 2, \cdots, n$, where n is the number of elements in the array, and $A[1] = A'[1] = B[1]$. The difference array derives from $B[i] = A'[i] - A[i]$. Therefore $A[i] = A'[i] - B[i] = (A[i-1] + 1) - B[i])$.

Observation

1. In Example 8.1, we predict that the value of each datum, from the second position, is its previous datum plus 1. We then compute the difference between the predicted $\mathbf{A}'$ and actual $\mathbf{A}$ values, scale it, and store it in an array. This array $\mathbf{B}$ is called a *residual array*.
2. The prediction 'rules' allow us to change a set of data to another with,

hopefully, a better distribution (for example, from a flat distribution to a more skewed one as in Figure 8.3). This is a *transform* in the sense that every datum in the original set has been changed. The transform is achieved by a simple mathematical formula.

3. Of course, this process has to be *reversible* in order to recover the original data during the decoding process.

8.1.1 Encoding

The prediction method we introduced in the previous section can be extended to a two-dimensional situation[1] in many applications.

Example 8.2 *Suppose that the following matrix* **A** *represents the pixel values of part of a larger greyscale image, where row indices $i = 1, \cdots, 8$ and column indices $j = 1, \cdots, 8$:*

$$\mathbf{A} = \begin{matrix} 2 & 4 & 2 & 3 & 1 & 1 & 1 & 1 \\ 1 & 3 & 2 & 3 & 5 & 1 & 1 & 1 \\ 3 & 4 & 5 & 5 & 5 & 5 & 5 & 5 \\ 6 & 4 & 5 & 8 & 7 & 9 & 4 & 5 \\ 3 & 2 & 7 & 3 & 2 & 7 & 9 & 4 \\ 3 & 3 & 4 & 3 & 4 & 4 & 2 & 2 \\ 1 & 2 & 1 & 2 & 3 & 3 & 3 & 3 \\ 1 & 1 & 1 & 2 & 2 & 3 & 3 & 3 \end{matrix}$$

Let us predict that from the second column each pixel is the same as the one to its left. So the residual matrix $R[i, j] = A[i, j] - A[i, j-1]$ is as below, where $i = 1, \cdots, 8$, $j = 2, \cdots, 8$:

$$\mathbf{R} = \begin{matrix} \mathbf{2} & 2 & -2 & 1 & -2 & 0 & 0 & 0 \\ \mathbf{1} & 2 & -1 & 1 & 2 & -4 & 0 & 0 \\ \mathbf{3} & 1 & 1 & 0 & 0 & 0 & 0 & 0 \\ \mathbf{6} & -2 & 1 & 3 & -1 & 2 & -5 & 1 \\ \mathbf{3} & -1 & 5 & -4 & -1 & 5 & 2 & -5 \\ \mathbf{3} & 0 & 1 & -1 & 1 & 0 & -2 & 0 \\ \mathbf{1} & 1 & -1 & 1 & 1 & 0 & 0 & 0 \\ \mathbf{1} & 0 & 0 & 1 & 0 & 1 & 0 & 0 \end{matrix}$$

We can now apply any methods of prefix coding, i.e. those text compression methods for generating a prefix code such as Huffman coding, to encode according to the frequency distribution of **R**.

The frequency distribution for this part of the residual image is then as below:

[1] Review Appendix B if necessary.

Entry	Occurrence	Probability
−5	2	2/64
−4	2	2/64
−2	4	4/64
−1	6	6/64
0	21	21/64
1	16	16/64
2	6	6/64
3	4	4/64
5	2	2/64
6	1	1/64

The entropy of the distribution is $-\sum_i p_i \log_2 p_i$ bits. The entropy of the entire image can be predicted as the *same* as that of this partial image. In other words, the entropy of a partial image can be used as an estimation for the entropy of the whole image. Of course, the compression quality of this approach depends very much on the difference between the real entropy and the estimated entropy of the entire image. Hopefully the overall compression would be a better one in comparison to the original 8 bits/pixel (e.g. ASCII) coding.

8.1.2 Decoding

The predictive transform process is reversible since the original matrix $A[i,j] = A[i,j-1] + R[i,j]$, where $A[i,0] = R[i,0]$ and $i = 1, \cdots, 8$, $j = 2, \cdots, 8$. Hence the whole predictive encoding process is lossless if the prefix encoding is lossless. The reader can easily justify this by completing Example 8.2 in the previous section.

8.2 Move to Front coding

Some prediction and transform methods require little mathematics. Move to Front (MtF) coding is a good example.

The idea of MtF is to encode a symbol with a '0' as long as it is a recently repeating symbol. In this way, if the source contains a long run of identical symbols, the run will be encoded as a long sequence of zeros.

Initially, the alphabet of the source is stored in an array and the index of each symbol in the array is used to encode a corresponding symbol. On each iteration, a new character is read and the symbol that has just been encoded is moved to the front of the array. This process can be seen easily from the example below.

Example 8.3 *Suppose that the following sequence of symbols is to be compressed:* `DDCBEEEFGGAA` *from a source alphabet* (`A, B, C, D, E, F, G`)*. Show how the MtF method works.*

Encoding

Initially, the alphabet is stored in an array:

```
0 1 2 3 4 5 6
A B C D E F G
```

1. Read `D`, the first symbol of the input sequence. Encode `D` by index 3 of the array, and then move `D` to the front of the array:

```
0 1 2 3 4 5 6
D A B C E F G
```

2. Read `D`. Encode `D` by its index 0, and leave the array unchanged because `D` is already at the front position of the array.

```
0 1 2 3 4 5 6
D A B C E F G
```

3. Read `C`. Encode `C` by its index 3, and move `C` to the front:

```
0 1 2 3 4 5 6
C D A B E F G
```

4. Read `B`. Encode it by 3 and move `B` to the front:

```
0 1 2 3 4 5 6
B C D A E F G
```

5. Read `E`. Encode it by 4 and move `E` to the front:

```
0 1 2 3 4 5 6
E B C D A F G
```

$\vdots$

and so on.

This process continues until the entire string is processed. Hence the encoding is 3, 0, 3, 3, 4, $\cdots$.

In this way, the more frequently occurring symbols are encoded by 0 or small decimal numbers.

Decoding

Read the following codes: 3, 0, 3, 3, 4, $\cdots$.

Initially,

```
0 1 2 3 4 5 6
A B C D E F G
```

1. Read 3, decode it to `D`, and move it to the front of the array:

```
0 1 2 3 4 5 6
D A B C E F G
```

2. Read 0, decode it to `D`, and do nothing since `D` is already at the front.

```
0 1 2 3 4 5 6
D A B C E F G
```

3. Read 3, decode it to `C`, and move it to the front.

```
0 1 2 3 4 5 6
C D A B E F G
```

$\vdots$

and so on.

In this way, the original sequence of symbols will be recovered one by one. We decode the first three symbols `DDC` $\cdots$ and decoding the entire sequence will be left to the reader as an exercise.

8.3 Burrows-Wheeler Transform (BWT)

The Burrows-Wheeler transform algorithm is the base of a recent powerful software program for conservative data compression `bzip` which is currently one of the best general purpose compression methods for text. The BWT algorithm was introduced by Burrows and Wheeler in 1994. The implementation of the method is simple and fast.

The encoding algorithm manipulates the symbols of **S**, the entire source sequence by changing the order of the symbols. The decoding process transforms the original source sequence back. During the encoding process, the entire input sequence of symbols is permutated and the new sequence contains hopefully some favourable features for compression.

In the example below, the encoding process produces a sequence **L** and an index s. The decoding algorithm reproduces the original source sequence back using **L** and an index s.

Example 8.4 *Consider a string* **S** *= 'ACCELERATE' of $n = 10$ characters, stored in a one-dimensional array. Show how BWT can be realised for encoding and decoding purposes.*

Encoding

The purpose of this process is to shuffle the symbols of the source sequence **S** in order to derive **L**, a new sequence which allows a better compression. The length of the original array **S** and of the resulting array **L** are the same because **L** is actually a permutation of **S**. In other words, we only want to change the order of the symbols in the original array **S** to get a new array **L** which can hopefully be compressed more efficiently.

Of course, the new array **L** cannot be just any array after a few random shuffles. There may be other ways to shuffle an array but the one used in BWT works well in practice.

Deriving L

In order to get **L**, we need to do the following steps:

1. We first shift the string **S** one symbol to the left in a circular way. By *circular*, we mean that the leftmost symbol in the array is shifted out of the array, and then added back from the right and becomes the rightmost element in the array. For example, '*A* C C E L E R A T E' will become 'C C E L E R A T E *A*' after such a *circular-to-left* shift.

```
(1)     0 1 2 3 4 5 6 7 8 9
      ← A C C E L E R A T E ←

(2)     0 1 2 3 4 5 6 7 8 9
      A C C E L E R A T E

(3)     0 1 2 3 4 5 6 7 8 9
        C C E L E R A T E A
```

Figure 8.4: A circular shift to the left

Repeating the circular shift $n-1$ times, we can generate the $n \times n$ matrix below where n is the number of symbols in the array, and each row and the column is a particular permutation of **S**.

	0	1	2	3	4	5	6	7	8	9
0	A	C	C	E	L	E	R	A	T	E
1	C	C	E	L	E	R	A	T	E	A
2	C	E	L	E	R	A	T	E	A	C
3	E	L	E	R	A	T	E	A	C	C
4	L	E	R	A	T	E	A	C	E	E
5	E	R	A	T	E	A	C	E	E	L
6	R	A	T	E	A	C	E	E	L	E
7	A	T	E	A	C	E	E	L	E	R
8	T	E	A	C	E	E	L	E	R	A
9	E	A	C	E	E	L	E	R	A	T

2. We now sort the rows of the matrix in lexicographic order so the matrix becomes:

	0	1	2	3	4	5	6	7	8	9
0	A	C	C	E	L	E	R	A	T	E
1	A	T	E	A	C	E	E	L	E	R
2	C	C	E	L	E	R	A	T	E	A
3	C	E	L	E	R	A	T	E	A	C
4	E	A	C	E	E	L	E	R	A	T
5	E	L	E	R	A	T	E	A	C	C
6	E	R	A	T	E	A	C	E	E	L
7	L	E	R	A	T	E	A	C	E	E
8	R	A	T	E	A	C	E	E	L	E
9	T	E	A	C	E	E	L	E	R	A

3. We name the last column **L**, which is what we need in the BWT for encoding, where s_1 indicates the first symbol of the given array **S**.

	F		**L**	
0	A	C C E L E R A T	E	
1	A	T E A C E E L E	R	
2	C	C E L E R A T E	A	$\leftarrow s_1$
3	C	E L E R A T E A	C	
4	E	A C E E L E R A	T	
5	E	L E R A T E A C	C	
6	E	R A T E A C E E	L	
7	L	E R A T E A C E	E	
8	R	A T E A C E E L	E	
9	T	E A C E E L E R	A	

Observation

1. **L** can be stored in a one-dimensional array, and can be written as

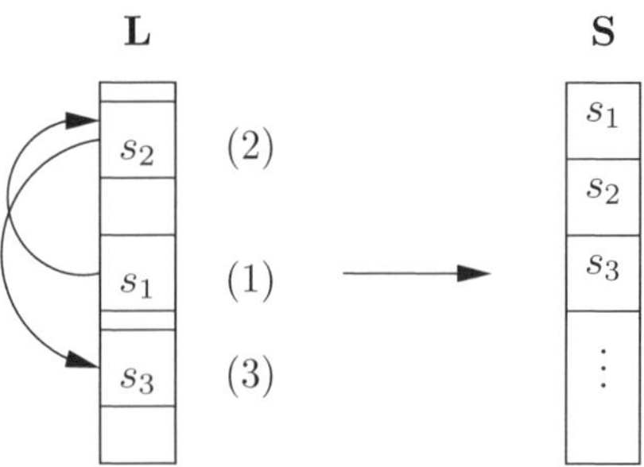

Figure 8.5: Finding the original string $\mathbf{S} = s_1 s_2 \cdots s_n$ from **L**

```
0 1 2 3 4 5 6 7 8 9
E R A C T C L E E A
```

2. Similarly, if we name the first column of the matrix **F**, then **F** can be derived from **L** by just sorting **L**. In other words, we should get **F** if we sort **L**.

   This is the sorted **L**, or **F**:

```
0 1 2 3 4 5 6 7 8 9
A A C C E E E L R T
```

3. As we will see later, the *first* symbol in **S** (i.e. 'A') is also transmitted in BWT to the decoder for decoding purposes.

Decoding

The goal of the reverse process of the transform is to recover the original string **S** from **L**. Since **L** is a permutation of **S**, what we need to find is the order relationship in which the symbols occurred in the original string **S**. Figure 8.5 shows the order relationship among the first three symbols in the original string $\mathbf{S} = s_1 s_2 s_3 \cdots$, where (1), (2) and (3) represent the order in which we shall find these symbols. An implicit chain relationship can be discovered among the elements in **L**.

We shall demonstrate how this implied chain relationship can be found in **L** with the help of its sorted version **F**, given the *first* character in the original **S**.

Note that both **L** and **F** are permutations of the original string **S**, and for each symbol in **L**, we know the next symbol in the original string **S** would be the one in **F** with the same index (because, during the encoding process, the leftmost element was shifted out and added to the right end to become the rightmost element). In other words, given any symbol in **L**, say the symbol s_i at k location, i.e. $L[k] = s_i$, we know that the next symbol in the original string **S**: $s_{i+1} = F[k]$. If we know where the s_{i+1} is in **L**, we would know the s_{i+2}. In this way, we can retrieve every symbol in **S**.

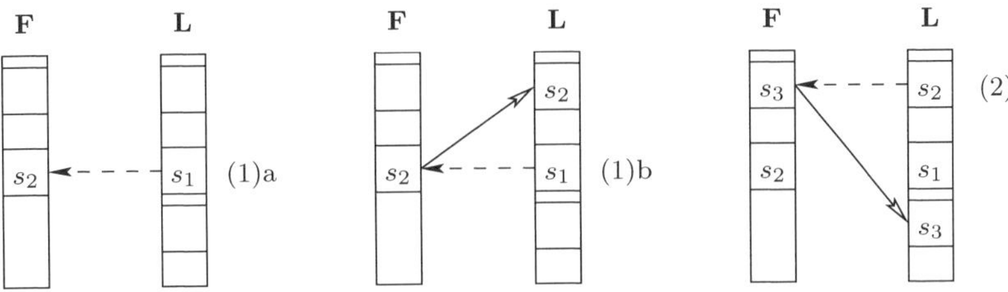

Figure 8.6: Chain relationship

In Figure 8.6(1)a, for example, since we know the position (index) of the first symbol s_1 in **L**, we can find the next symbol s_2 in **F** using the same index. This relationship is illustrated by a dashed arrow from s_1 in **L** to s_2 at the same position in **F**.

In Figure 8.6(1)b, s_2 can be identified in **L** using an auxiliary array (see later). A solid arrow represents the position (index) link between the different locations of the same symbol in **F** and in **L**.

Note that two types of links exist between the items in **F** and those in **L**. The first type is, from **L** to **F**, to link two items s_i and its follower s_{i+1} by an identical index. If $s_i = L[k]$, for some index k, then the next symbol $s_{i+1} = F[k]$. The second type is, from **F** to **L**, to link by the same symbol s_{i+1} from its location in **F** to its location in **L**.

Figure 8.6(2) shows the two types of links **F** and **L**. The first type of link is represented as a dashed line with arrow from s_2 in **L** to its follower s_3 in **F**, and the second type as a solid line pointing from s_3's position in **F** to its position in **L**.

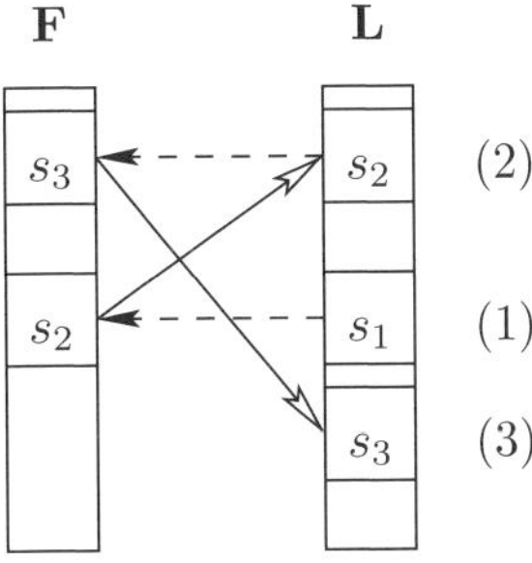

Figure 8.7: Finding the original string $\mathbf{S} = s_1 s_2 \cdots s_n$ from **L** via **F**

Figure 8.7 shows how, by following the two links in Figure 8.6 on each step, we can retrieve the original sequence of symbols one after another, for example from s_1 to s_2 and then from s_2 to s_3 and so on.

The first type of link can be built up straight away using an identical index each time. The second type of link, however, requires knowing the location in **L** for each symbol in **F**. For example, as in Figure 8.6(1)b, before finding the follower of s_2, we need to know the index k of s_2 in **L**.

Fortunately, this information is easy to establish when we derive **F** from **L**. We only need to store the index of each of the symbols in **L** while sorting them.

For example, we can define an auxiliary array **T** to store, for each element in **F**, its index in **L**. **T** is sometimes called *transformation vector* and is critical for deriving the original string **S** from **L**. Alternatively, we can define the items in **F** as an array of `objects` and store the symbols and their index information in **L** in different fields.

Example 8.5 *Consider the* **L**, **F** *and auxiliary array* **T**. *Suppose the first symbol* s_1 *is* $L[2] = A$ *(shade in Figure 8.8). Show with the two types of links how the original string* `A C C E L E R A T E` *can be derived.*

```
   0 1 2 3 4 5 6 7 8 9
L: E R A C T C L E E A

   0 1 2 3 4 5 6 7 8 9
F: A A C C E E E L R T
T: 2 9 3 5 0 7 8 6 1 4
```

Starting from the first symbol in $L[2] = s_1 =$ '`A`' which is known (transformed from the encoder). Following the links in Figure 8.8, we can easily derive the original string **S**.

The process starts from $s_1 = L[2]$, we then know the next symbol in **S**: $s_2 = F[2] =$ '`C`' $= L[T[2]] = L[3]$ (Figure 8.8(1)). From $s_2 = L[3]$, we know the next symbol is $s_3 = F[3] =$ '`C`' $= L[T[3]] = L[5]$ (Figure 8.8(2)). From $s_3 = L[5]$, we know the next symbol $s_4 = F[5] =$ '`E`' $= L[T[5]] = L[7]$. This process continues until the entire string **S** = `ACCELERATE` is derived (Figure 8.8(10)).

Observation

1. The BWT algorithm can be fast using array implementation.
2. The BWT algorithm can perform well in certain situations as a preprocessing compression algorithm and works best with other standard compression algorithms such as RLE, Huffman etc.
3. The implementation of the algorithms is fairly simple.
4. The BWT cannot be performed until the entire input file has been processed.
5. There is a typical preprocess by RLE and post-process of Huffman coding.

8.4 Transform approach

Transform is a standard mathematical tool being used in many areas to solve sometimes difficult computation problems in the original form. The main idea is to change a group of quantity data such as a vector or a function to another form in which some useful features may occur. For example, the computation required in the new form may become more feasible in the new space. The

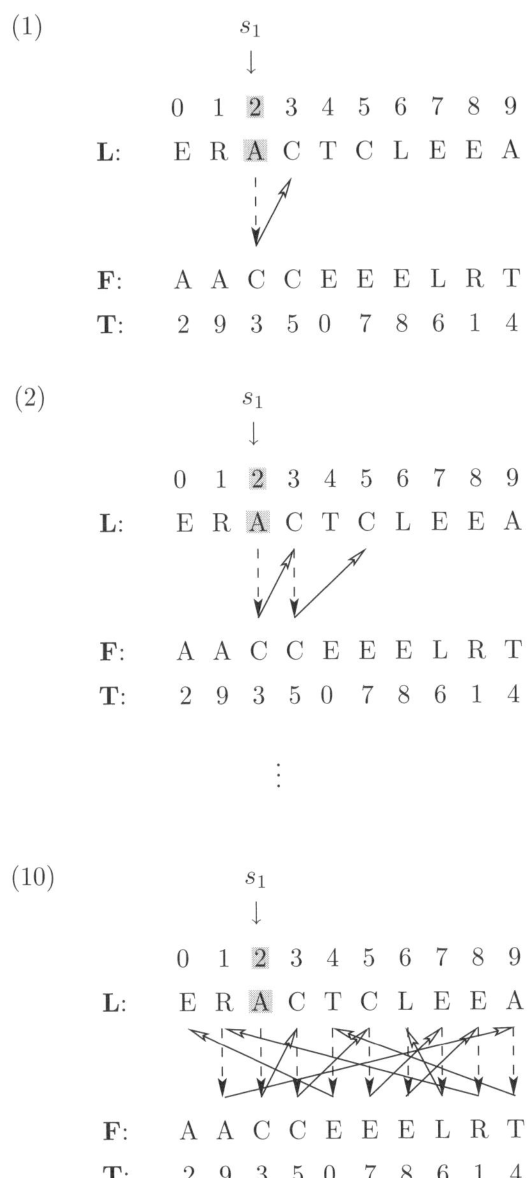

Figure 8.8: Finding the original string **S** from **L**

computation result in the new form is then converted back to the original form by the *inverse* transform.

The *transform* approach changes source data domain. The process of changing from the current domain to the new domain is called *forward transform* (*transform* for short) and the process of changing back to the original domain is called *inverse transform*. We first look at an example of a simple transform below:

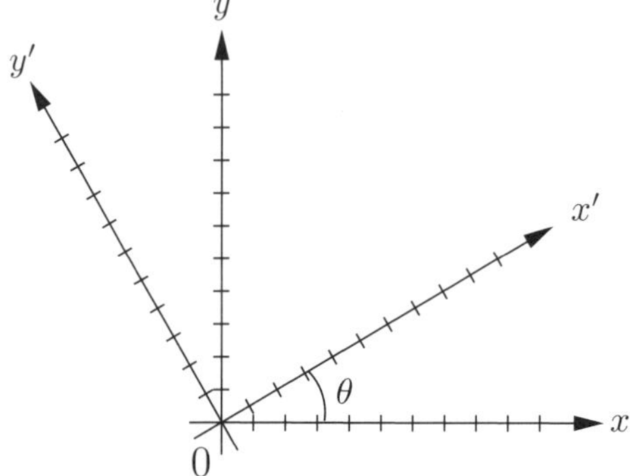

Figure 8.9: From x-y domain to x'-y' domain

Example 8.6 *Figure 8.9 shows a simple geometric rotation of angle θ, from space x-y to space x'-y', which can be achieved by applying a transform matrix* $\mathbf{T}$ *below (see Appendix B for background information on matrices).*

Each geometric point in x-y space can be represented by a pair of $v = (x_i, y_i)$. The rotation transformation matrix can be defined as

$$\mathbf{T} = \begin{pmatrix} \cos\theta & \sin\theta \\ -\sin\theta & \cos\theta \end{pmatrix}$$

Each new point $v' = (x'_i, y'_i)$ in the new x'-y' space can then be derived by the following matrix computation:

$$\mathbf{v}' = \mathbf{T}\mathbf{v} = \begin{pmatrix} \cos\theta & \sin\theta \\ -\sin\theta & \cos\theta \end{pmatrix} \begin{pmatrix} x \\ y \end{pmatrix}$$

The inverse transformation matrix T^{-1} is

$$\mathbf{T}^{-1} = \begin{pmatrix} \cos\theta & -\sin\theta \\ \sin\theta & \cos\theta \end{pmatrix}$$

The formula $\mathbf{v}' = \mathbf{T}\mathbf{v}$ is called a *forward transform*, or *transform* for short, and $\mathbf{v} = \mathbf{T}^{-1}\mathbf{v}'$ is called *inverse transform*, where $\mathbf{T}$ is called the *transform matrix*.

If the geometric object in the original x-y space turns out to be difficult, then a suitable transform may provide an easier way to look at the geographic object. Example 8.7 shows what happens with a set of data after a rotation transform.

Example 8.7 *Consider the original data pair:*
$\mathbf{x} = (7, 16, 37, 34, 20, 27, 40, 10, 31, 16, 22, 16, 43, 3, 39)$;
$\mathbf{y} = (20, 34, 69, 59, 41, 50, 68, 13, 51, 20, 36, 27, 76, 12, 66)$ *as in Figure 8.10. Show how the set of 30 data (15 x values and 15 y values) can be transformed after a $\pi/3$ degree clockwise rotation.*

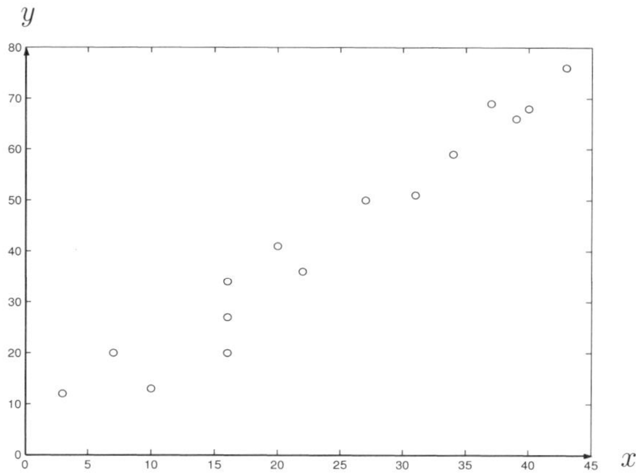

Figure 8.10: Before transform

Since the data concentrate along a line with an angle of $\theta = \pi/3$ degrees against the x axis, we define the transform function as follows:

$$\mathbf{T} = \begin{pmatrix} \cos(\pi/3) & \sin(\pi/3) \\ -\sin(\pi/3) & \cos(\pi/3) \end{pmatrix} = \begin{pmatrix} 0.5000 & 0.8660 \\ -0.8660 & 0.5000 \end{pmatrix}$$

We obtain the following data which are rounded to integer for convenience (Figure 8.11):
$\mathbf{x}' = (21, 37, 78, 68, 46, 57, 79, 16, 60, 25, 42, 31, 87, 12, 77)$;
$\mathbf{y}' = (4, 3, 2, 0, 3, 2, -1, -2, -1, -4, -1, 0, 1, 3, -1)$.

As we can see the data points now concentrate along the x' axis after the transform. Now an entropy compression algorithm can be applied to the data in the x'-y' domain.

In fact, the transform has a more interesting impact on the data. Figure 8.12(1) shows the data distribution in the x-y domain and Figure 8.12(2) shows the data distribution in the x'-y' domain after the transformation. The small circle represents the value of an x and the small star the value of a y. As we can see, while some x's' values increase, most y's' values in the x'-y' domain decrease and become very close to zero. If this is an application that allows lossy compression, even higher compression may be achieved.

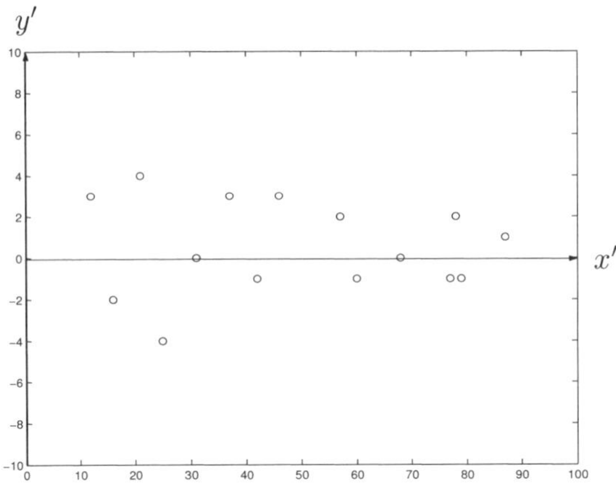

Figure 8.11: After transform

The reverse transform can be easily completed by applying

$$\mathbf{T}^{-1} = \begin{pmatrix} \cos(\pi/3) & -\sin(\pi/3) \\ \sin(\pi/3) & \cos(\pi/3) \end{pmatrix} = \begin{pmatrix} 0.5000 & -0.8660 \\ 0.8660 & 0.5000 \end{pmatrix},$$

and setting $y'_i = 0, i = 1, \cdots, 15$, we can reconstruct a set of approximate data:

$\mathbf{x_e} = (11, 19, 39, 34, 23, 29, 40, 8, 30, 13, 21, 16, 44, 6, 39)$;
$\mathbf{y_e} = (18, 32, 68, 59, 40, 49, 68, 14, 52, 22, 36, 27, 75, 10, 67)$.

Observation

1. Data compression can be achieved only by storing x's' values.
2. The energy of the original data can be computed by

$$E = \sum_{i=1}^{n} ({x_i}^2 + {y_i}^2) = 45\,069$$

   and the energy of the transformed data

$$E' = \sum_{i=1}^{n} ({x'_i}^2 + {y'_i}^2) = 45\,069$$

   This means that the energy remains the same before and after the transform. However, the energy is concentrated in the x's only among the transformed data.
3. The reconstructed set of data is slightly different from the original ones since we set $y'_i = 0$. This means that the whole compression process is lossy.

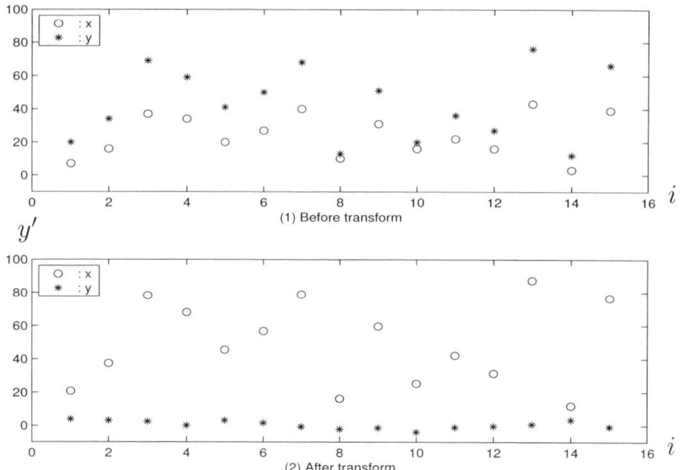

Figure 8.12: Comparison of data distribution before and after transform

4. The loss is caused by rounding the real data. Rounding the data is the simplest form of quantisation (Section 9.3). The difference between the reconstructed data and the original data is called *distortion*. We shall look at the issues in these lossy operations in the next chapter.

8.4.1 Orthogonal transform

We have observed from Example 8.7 that the transform can make data more skewed by increasing some of the values and decreasing others. By further checking on the total of $x^2 + y^2$, we find the exact same amount of $\sum_{i=1}^{n}({x_i}^2 + {y_i}^2) = 45\,069$ in both the x-y domain and the x'-y' domain. In other words, the transform does not change this total amount of energy in both domains.

It turns out that this is not a coincidence. The amount $\sum_{i=1}^{n}({x_i}^2 + {y_i}^2)$ is called *energy* of the system. The transform we have used is a so-called *orthogonal transform*, a favourable type for data compression, used also in many applications in other fields.

A transform is orthogonal if the transform matrix is orthogonal (Appendix B.4). One immediate advantage of applying an orthogonal transform is that the inverse transform can be easily found. If the transform matrix is $\mathbf{T}$, then the reverse transform matrix is just $\mathbf{T}^T$, the transpose of $\mathbf{T}$.

However, the most important advantage of applying an orthogonal transform is that first, such a transform keeps the *energy* of the data (defined as the sum of the square of each datum) unchanged. Secondly, in the transformed vector, the first few elements often concentrate a large proportion of the system energy. Hence these few elements can be used to give a good approximation of the entire original set of data. In other words, these few elements only may be sufficient for reconstructing the original set of data and the rest of transformed values can be dropped. Therefore, an orthogonal transform can be applied to achieve good data compression.

We look at an example below.

Let $\mathbf{x}'$ and $\mathbf{x}$ be vectors and $\mathbf{T}$ be an $n \times n$ matrix. Suppose x'_i, x_i and t_{ij} are the elements in $\mathbf{x}'$, $\mathbf{x}$ and $\mathbf{T}$ respectively. Let $x'_i = \sum_j t_{ij}x_j$, where x'_i is a weighted sum of x_j and is multiplied by a weight t_{ij}, $i, j = 1, \cdots, 3$.

This can be written in matrix form:

$$\begin{pmatrix} x'_1 \\ x'_2 \\ x'_3 \end{pmatrix} = \begin{pmatrix} t_{11} & t_{12} & t_{13} \\ t_{21} & t_{22} & t_{23} \\ t_{31} & t_{32} & t_{33} \end{pmatrix} \begin{pmatrix} x_1 \\ x_2 \\ x_3 \end{pmatrix}$$

We can write $\mathbf{x}' = \mathbf{T}\mathbf{x}$. Each row of $\mathbf{T}$ is called the *basis vector*.

In practice, the weights should be independent of the values of x_is so the weights do not have to be included in the compressed file.

Example 8.8 *Given a weight matrix*

$$\mathbf{T} = \frac{1}{2}\begin{pmatrix} 1 & 1 & 1 \\ 1 & -1 & -1 \\ 1 & -1 & 1 \end{pmatrix}$$

and the transform vector

$$\mathbf{x} = \begin{pmatrix} 4 \\ 6 \\ 5 \end{pmatrix}$$

we have

$$\frac{1}{2}\begin{pmatrix} 1 & 1 & 1 \\ 1 & -1 & -1 \\ 1 & -1 & 1 \end{pmatrix}\begin{pmatrix} 4 \\ 6 \\ 5 \end{pmatrix} = \frac{1}{2}\begin{pmatrix} 15 \\ -7 \\ 3 \end{pmatrix} = \begin{pmatrix} 7.5 \\ -3.5 \\ 1.5 \end{pmatrix}$$

Let us calculate the energy after and before the transform. The energy of $\mathbf{x}' = (7.5, -3.5, 1.5)^T$ and that of $(4, 6, 5)^T$ are $7.5^2 + (-3.5)^2 + 1.5^2 = 70.75$ and $4^2 + 6^2 + 5^2 = 77$ respectively. Comparing the two, we find that most of the energy is conserved during the transform. However, the energy in each individual element in the transformed vector is $(7.5^2, (-3.5)^2, 1.5^2) = (56.25, 12.25, 2.25)$. Most energy is concentrated in 7.5, the first element of $\mathbf{x}'$. This suggests a possible compression even in this small example, because we can store, instead of the entire $\mathbf{x}'$, the first element (7.5) only, and remove the other elements (-3.5 and 1.5).

When reconstructing the original $\mathbf{x}$, we first quantify the transformed vector $\mathbf{x}'$ from $(7.5, -3.5, 1.5)^T$ to integers $(8, -4, 1)^T$ and conduct an inverse transform: $\mathbf{x} = \mathbf{T}^{-1} \cdot \mathbf{x}'$

$$(1/2)\begin{pmatrix} 1 & 1 & 1 \\ 1 & -1 & -1 \\ 1 & -1 & 1 \end{pmatrix}\begin{pmatrix} 8 \\ -4 \\ 2 \end{pmatrix} = \begin{pmatrix} 3 \\ 5 \\ 7 \end{pmatrix}$$

If $(8,0,0)^T$ is used to reconstruct $\mathbf{x}$, we have

$$(1/2)\begin{pmatrix} 1 & 1 & 1 \\ 1 & -1 & -1 \\ 1 & -1 & 1 \end{pmatrix}\begin{pmatrix} 8 \\ 0 \\ 0 \end{pmatrix} = \begin{pmatrix} 4 \\ 4 \\ 4 \end{pmatrix}$$

In practice, the $\mathbf{T}$ is much bigger so the reconstruction result would be better. More sophisticated transform techniques, such as DCT and DWT (see following sections), would produce better results.

8.5 Discrete Cosine Transform (DCT)

A periodic function can be represented as the sum of sines and cosines. This fact was discovered by the French mathematician and physicist Jean Baptiste Joseph Fourier (1768–1830) and the work was published in 1822. It has been regarded as one of the most revolutionary contributions of the nineteenth century. His theorem was later substantiated by Peter Gustav Lejeune Dirichlet (1805–59).

The fact that all periodic functions can be expressed as a sum of sinusoidal components suggests a useful approximation of the function because the first few terms of the infinite series are normally sufficient for many applications. It also provides an analysis tool for decomposition of a compound waveform.

The Discrete Cosine Transform, similar to the Fourier transform (see Appendix C), represents a signal by its elementary frequency components. DCT relates to the Discrete Fourier Transform (DFT) closely and this can be seen from the DFT formula. However, it performs much better than DFT for compression. DCT has a wide range of applications in data compression as well as in other subject areas.

The commonly used Discrete Cosine Transform is two-dimensional and can be described by an $n \times n$ transform matrix C as below. Each entry of the matrix $C[i,j]$ is obtained from a function of cosines:

$$C[i,j] = \begin{cases} \sqrt{\frac{1}{n}}\cos\frac{(2j+1)i\pi}{2n}, & i = 0;\ j = 0, 1, \cdots, n-1 \\ \sqrt{\frac{2}{n}}\cos\frac{(2j+1)i\pi}{2n}, & i = 1, \cdots, n;\ j = 0, 1, \cdots, n-1 \end{cases}$$

The main advantages of DCT compared to other transforms are that DCT does not introduce any sharp discontinuities at the edges, and it works substantially better in energy compaction for most correlated sources. DCT was included in JPEG and MPEG in the years prior to JPEG 2000 where wavelet transform methods are included.

There are other common transforms with similar approaches such as the Discrete Walsh-Hadamand Transform (DWHT) and the Discrete Karhunen-Loeve Transform (KLT). DWHT requires mostly additions and subtractions so it has good computational efficiency although it does not give a good effect on continuous data. KLT can achieve the optimal energy concentration but may require expensive computation because the transform matrix is required to be calculated for each individual set of data.

8.6 Subband coding

The performance of the transform techniques that we have seen so far depends very much on certain well-defined characteristics of the source data. Hence none of these transforms would work very well on its own if the source contains a combination of two or more controversial characteristics. In this case, certain algorithms are needed to decompose the source data into different frequency components first to highlight individual characteristics. The components are then encoded separately. In the reconstruction process, the components are decoded before being assembled to recover the original signal. Such encoding and decoding processes are called *subband coding.*

The algorithms that isolate certain frequency components in a signal are called *filters.* The signal is decomposed into several frequency components which are called *subbands.* Filters that allow certain subbands to pass are called *band-pass* filters.

A *low-pass* filter allows the low frequency components to pass but blocks the high frequency components. The threshold frequency f_0 is called the *cutoff* frequency. Similarly, a *high-pass* filter allows the subbands above the cutoff frequency to pass but blocks those below.

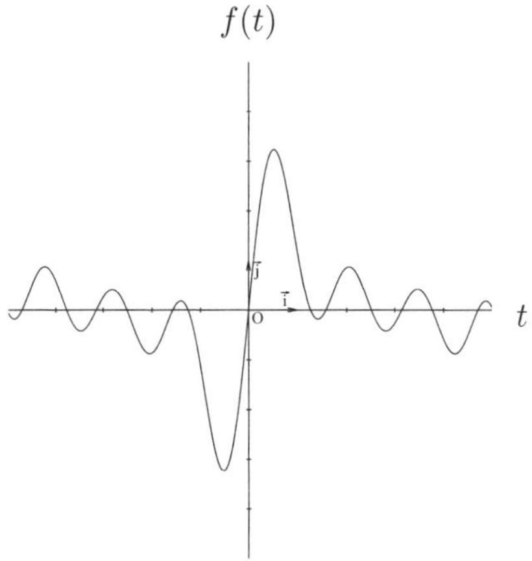

Figure 8.13: A rapidly changing source

With proper filters, the combination characters may be decomposed and certain characteristics may be identified and selected. Figure 8.13 shows a source appears changing rapidly. However, it contains a low frequency component (Figure 8.14), which can be extracted using a low-pass filter.

A general purpose filter can be modelled using the mathematical formula below:

$$y_n = \sum_{i=0}^{N} a_i x_{n-i} + \sum_{i=1}^{M} b_i y_{n-i},$$

where the sequence (x_n) is the input to the filter, the sequence (y_n) is the output from the filter, and (a_i) and (b_i) are the *filter coefficients.*

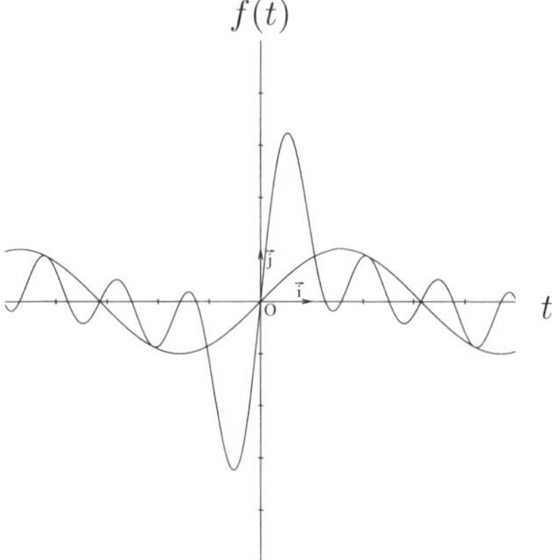

Figure 8.14: Showing a slow movement of $\sin(x)$

A number of filters may be used one after another or in parallel to achieve better decomposition. A collection of filters is often called *filter banks*. Filters used in the decoding process are usually called *analysis banks* and those used in the decoding process are *synthesis banks*. Various filter banks for subband coding can be found in the literature or in off-the-shelf programs. Topics on the selection, analysis or design of the filter banks are beyond the scope of this book. Interested readers are encouraged to consult the literature.

8.7 Wavelet transforms

A wave can be defined as a function of time, $f(t)$, where $(-\infty < t < \infty)$. Waves of specific characteristics with a *limited* duration are called *wavelets*.

Wavelets are mathematical functions that satisfy certain conditions. Like $\sin(t)$ and $\cos(t)$ in Fourier transform, wavelets can be used to represent data or other functions. However, wavelets methods have advantages over Fourier methods in analysing signals containing spikes or discontinuities. Wavelets were developed independently in different scientific fields such as mathematics, quantum physics, and electrical engineering.

In the area of data compression, wavelet transforms allow a similar transform in the DFT, i.e. only storing the main coefficients of series of basis functions and the transform matrices.

Wavelet transforms gained their popularity in recent years due to their good performance and being included in JPEG 2000. The two-dimensional wavelet transform technique can be used for image compression.

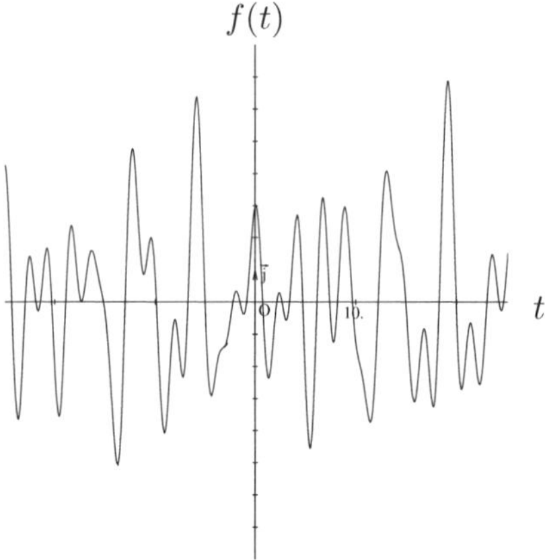

Figure 8.15: A signal whose frequency varies according to time

8.7.1 Scale-varying basis functions

A scale-varying basis function is a basis function that varies in scale. It is also called a *step-function*. The function or data space is divided using different scale size. For example, if we have a function $f(t)$ with a range of [0, 1], we can divide the function into two step-functions:

$$f(t) = \begin{cases} f_1(t) & 0 < t \leq 1/2, \\ f_2(t) & 1/2 < t \leq 1. \end{cases}$$

We can then further divide the functions to four step-functions:

$$f(t) = \begin{cases} f_{11}(t) & 0 < t \leq 1/4, \\ f_{12}(t) & 1/4 < t \leq 1/2, \\ f_{21}(t) & 1/2 < t \leq 3/4, \\ f_{22}(t) & 3/4 < t \leq 1. \end{cases}$$

Let $g_i(t)$ be another set of functions known as the *basis*. Suppose a function f can be represented as the weighted sum of some set of $g_i(t)$:

$$f(t) \approx \sum_{i}^{N} \alpha_i g_i(t)$$

The values of α_i are called the *coefficients*, which can be Boolean, integer or real numbers. The compression process aims to find a good set of coefficients $(\alpha_1, \alpha_2, \cdots, \alpha_N)$ to represent signal f. The decompression process aims to reconstruct f from the set of coefficients. The compression and decompression algorithms share the same set of functions and the compressed file consists of the set of coefficients.

Like Fourier methods such as Fast Fourier Transform (FFT), wavelet transform is also a linear orthogonal transform. Hence the inverse transform matrix for Discrete Wavelet Transform (DWT) is also the transpose of the original. It can also be viewed as a 'rotation' in function space to a different domain. For DFT, the new domain contains basis functions of sine and cosine waves. For DWT, the new domain contains more complicated basis functions called wavelets, mother wavelets, or analysing wavelets. Also, the energy of the original function or data set is conserved after the transformation. The energy is often concentrated in the first few elements in the transformed vector (if it is the one-dimensional case). Hence only the few coefficients need to be encoded and very effective compression may be achieved.

The wavelet transforms work by choosing a set of representative patterns and finding a subset of these patterns that add up to the signal. It is essentially a subband transform. Note that wavelet transforms have an *infinite* set of possible basis functions. Unlike DFT, wavelet transforms do not have a single set of basis functions like the Fourier transform where only sine and cosine functions are involved.

The main issues in a wavelet transform in practice are:

- finding a good set of basis functions for a particular class of signal
- finding an effective quantisation method for different frequencies.

The wavelet approach has developed well in recent years. Research has been done in fast wavelet transform, on wavelet packets transform, and adaptive waveforms. The results offer many potential application areas. More well-defined and studied techniques are expected.

Summary

Prediction and transform approaches represent another generation of compression technologies. In order to achieve better compression ratio, different compression methods are used one after another. Preprocessing prepares a better data source for specific compression algorithms at a later stage. MtF and BWT algorithms are examples of lossless approach and DFT, DOC and wavelets are those of lossy. JPEG and MPEG offer standards and frameworks.

Learning outcomes

On completion of your studies in this chapter, you should be able to:

- explain the concept of prediction and transforms
- describe the main approaches in prediction and transforms
- outline some simple preprocessing algorithms such as MtF algorithms, BWT, DCT and DWT.

Exercises

E8.1 Derive a predictive rule, for example 'add 2 and store the difference', for $\mathbf{A} = (1, 3, 5, 6, 7, 9, 9, 11, 14)$ so that the residual array would have more skewed frequency distribution.

E8.2 Compute the entropies for $\mathbf{A} = (1, 3, 5, 6, 7, 9, 9, 11, 14)$ and its residual array in the above question.

E8.3 Suppose that the matrix **A** below represents the pixel values of part of a large image, where $i = 0, \cdots, 7$, $j = 0, \cdots, 7$, and $A[0, 0] = 4$. Let us predict that each pixel is the same as the one to its right.

(a) Derive the residual matrix **R** of **A**
(b) Derive the frequency table for **R**
(c) Derive a variable length code, e.g. Huffman code, for the partial image
(d) Discuss under what conditions the code can be applied to the whole image with less distortion.

$$\mathbf{A} = \begin{matrix} 4 & 8 & 4 & 8 & 1 & 1 & 1 & 1 \\ 1 & 2 & 4 & 6 & 5 & 1 & 1 & 1 \\ 8 & 4 & 5 & 5 & 5 & 5 & 5 & 5 \\ 2 & 4 & 8 & 5 & 7 & 9 & 5 & 5 \\ 2 & 4 & 6 & 7 & 7 & 7 & 9 & 9 \\ 2 & 2 & 2 & 3 & 4 & 9 & 7 & 3 \\ 3 & 3 & 6 & 6 & 6 & 7 & 7 & 7 \\ 7 & 7 & 7 & 7 & 6 & 7 & 8 & 8 \end{matrix}$$

E8.4 Given the orthogonal matrix

$$\mathbf{W} = \begin{pmatrix} 1 & 1 & 1 & 1 \\ 1 & 1 & -1 & -1 \\ 1 & -1 & -1 & 1 \\ 1 & -1 & 1 & -1 \end{pmatrix}$$

and a data vector

$$\mathbf{D} = \begin{pmatrix} 4 \\ 6 \\ 5 \\ 2 \end{pmatrix},$$

conduct a similar experiment to the approach in this chapter.

(a) Fulfil a linear transform on **D**
(b) Perform a reverse transform to get $\mathbf{D}'$
(c) Compare $\mathbf{D}'$ with the original **D**.

E8.5 Implement the MtF algorithms and demonstrate how the encoding and decoding algorithms work on an example string `AABBBABABADDCC`.

Laboratory

L8.1 Design and implement a method which takes a matrix of integers and returns its residual matrix, assuming that each entry is the same as the one to its left plus 1.

L8.2 Design and implement a method which takes a residual matrix derived from the above question, and returns its original matrix.

L8.3 Design and implement the encoding and decoding algorithms for the MtF approach.

L8.4 Implement Algorithms 8.1 and 8.2 ([WMB99]) below.
Hint: you should work out the input, output and the meaning of each variable used first.

Algorithm 8.1 BWT encoding

INPUT: ?
OUTPUT: ?

sort the N input characters using the preceding characters as the sort key, create permuted array $P[1 \cdots N]$
output the position in P that contains the first character from the compressed file
output the permuted array P

Algorithm 8.2 BWT decoding

INPUT: ?
OUTPUT: ?

$p \leftarrow$ the position of the first input character (from the encoder)
$P[1 \cdots N] \leftarrow$ the permuted symbols (from the encoder)
$K[s] \leftarrow$ the number of times symbol s occurs in P
set the array $M[s]$ to be the position of the first occurrence of s in the array that would be obtained by sorting P
  (a) set the array M[first symbol in lexical order] $\leftarrow 1$
  (b) for (each symbol s (in lexical order)) $M[s] \leftarrow M[s-1]+K[s-1]$; end for
for $i = 1; i \leq N; i = i + 1$ **do**
  $s \leftarrow P[i]$; $L[i] \leftarrow M[s]$; $M[s] \leftarrow M[s] + 1$
end for
{Array L now stores the links with which to traverse the permuted string}
traverse the link array to reconstruct the original string
  (a) $i \leftarrow p$ (initial position)
  (b) for $(k = 1; k \leq N; k = k + 1)$ output $P[i]$; $i \leftarrow L[i]$; end for

---

L8.5 Design and implement your own version of the encoding and decoding algorithms for the BWT approach.

Assessment

S8.1 Describe the encoding and decoding algorithms in DCT and DWT.

S8.2 Show step by step how the BWT algorithm works on the transform and inverse transform of string `BBCADDBB`.

Bibliography

[BGG98] C.S. Burrus, R.A. Gopinath, and H. Guo. *Introduction to Wavelets and Wavelet Transforms.* Prentice Hall, Englewood Cliffs, New Jersey, 1998.

[BKS99] B. Balkenhol, S. Kurtz, and Y.M. Shtarkov. Modifications of the Burrows and Wheeler data compression algorithm. *In Storer and Cohn*, pages 188–197, 1999.

[BW94] M. Burrows and D.J. Wheeler. A block-sorting lossless data compression algorithm. Technical Report SRC 124, Digital Equipment Corporation, Palo Alto, California, May 1994.

[CT97] J.G. Cleary and W.J. Teahan. Unbounded length contexts for PPM. *The Computer Journal*, 40(2/3):67–75, February 1997.

[CW84] J.G. Cleary and I.H. Witten. Data compression using adaptive coding and partial string matching. *IEEE Transactions on Communications*, COM-32(4):396–402, April 1984.

[Dau92] I. Daubechies. *Ten Lectures on Wavelets.* SIAM, Philadelphia, PA, 1992.

[Deo00] S. Deorowicz. Improvements to Burrows-Wheeler compression algorithm. *Software-Practice and Experience*, 30(13):1465–1483, November 2000.

[EVKV02] M. Effros, K. Visweswariah, S. Kulkarni, and S. Verdu. Universal lossless source coding with the Burrows Wheeler transform. *IEEE Transactions on Information Theory*, 48(5):1061–1081, May 2002.

[WM01] I. Wirth and A. Moffat. Can we do without ranks in Burrows Wheeler transform compression? In *IEEE Data Compression Conference*, pages 419–428, Snowbird, Utah, March 2001.

[WMB99] I.H. Witten, A. Moffat, and T.C. Bell. *Managing Gigabytes: Compressing and Indexing Documents and Images.* Morgan Kaufmann Series in Multimedia Information and Systems. Morgan Kaufmann, 2nd edition, 1999.

Chapter 9

Audio compression

Audio compression has interested researchers from its early years but has become a hot topic since the 1990s due to the popular MP3 music, DVD, digital radio and digital TV technology. In this chapter, we discuss the fundamentals of compression techniques for audio data.

9.1 Modelling sound

Sound is essentially a sensation of the human audio system. It is detected by the ears and interpreted by the brain in a certain way.

Sound can be modelled as waves and described in two ways using mathematical functions. One is $s(t)$ in the time domain and the other $S(f)$ in the frequency domain, where t represents time and f represents the frequency.

The first sound wave was scribed by Leon Scott using a stiff bristle attached to a diaphragm actuated by a horn in 1857 [Tre78]. Koenig improved Scott's invention during 1858 to 1862 and presented his results in London in 1862 in 'Phonograms'.

In 1711, John Shore discovered the tuning fork, and in 1908, G. W. Pierce described in a paper, 'A simple method of measuring the intensity of sound' based on his work in measuring sound intensities in auditoriums and of train whistles; thus was born the first sound-level meter.

Sound, in a way similar to colour, is understood as a mixture of physical and psychological factors. We know that sound is a physical disturbance in a medium and it is propagated in the medium as a pressure wave by the movement of atoms or molecules. Hence sound is often described as a function $s(t)$ (as for electronic signals), measuring the pressure of medium at a time t. If T is the period measured in hertz (Hz), a sine wave can be written as $s(t) = s(t+T)$ and $\sin(2\pi t) = \sin(\frac{2\pi}{T}t)$ since $f = \frac{1}{T}$. A periodic signal $s(t)$ with period T can be represented by the sum of sine or cosine waves:

$$s(t) = \frac{a_0}{2} + \sum_{i=1}^{\infty} a_i \cos(i\frac{2\pi}{T}t) + \sum_{i=1}^{\infty} a_i \sin(i\frac{2\pi}{T}t)$$

Figure 9.1 shows some sine waves with different frequencies.

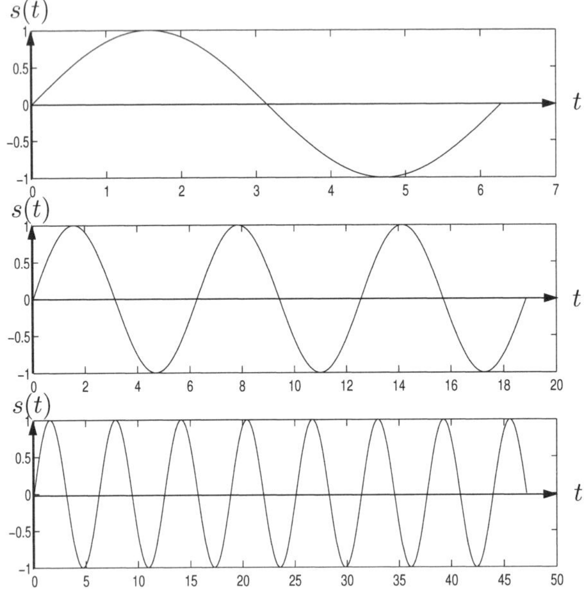

Figure 9.1: Three sine waves with different frequencies

Example 9.1 *Most orchestras tune the A above middle C note to 440 Hz. The sound pressure function for that note can be written therefore as* $s(t) = \sin(880\pi t)$ *(Figure 9.2).*

We also know that the human ear is normally able to detect frequencies in the range between 20 Hz and 20 kHz. The upper audible limit in terms of frequency tends to decrease as age increases. Children may hear sounds with frequencies as high as 20 kHz but few adults can.

It is interesting to notice that the timbre of different instruments is created by the transient behaviour of notes. In fact, if the attack portion is removed from recordings of an oboe, violin, and soprano creating the same note, the steady portions are indistinguishable.

Like any other wave functions, we can plot any sound signal by plotting its amplitude (pressure on medium) against time (see Figure 9.3 for an example). The shape of the curve can demonstrate clearly the different properties of a specific sound.

Alternatively, we can use a *frequency spectrum diagram* to represent the changes of frequencies over a period of time. In a frequency spectrum diagram, we plot the amplitude against the frequency.

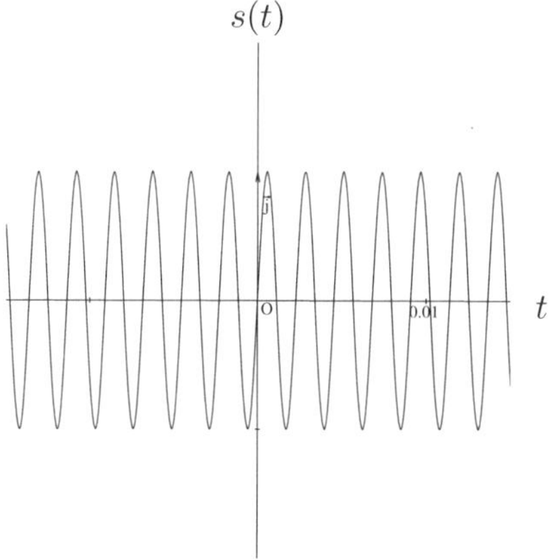

Figure 9.2: Plot $s(t) = \sin(880\pi t)$

Example 9.2 *Figure 9.4 is a frequency spectrum diagram for* $s(t) = 0.1\sin(2\pi t) + 0.3\sin(8\pi t) - 0.25\sin(20\pi t)$ *in Figure 9.3.*

The frequency spectrum can be easily measured and transformed. Any audio signal, in theory, can be reconstructed by the frequencies in the frequency spectrum. Hence the frequency spectrum instead of the signals themselves can be used to encode the audio signals. This technique was used first by Homer Dudley, an engineer at Bell Laboratories, in his invention of the first voice coder (called *vocoder*) in 1928. The speech signal with a bandwidth of over 3000 Hz was compressed into the 100 Hz bandwidth transatlantic telegraph cable. The vocoder is regarded as the grandfather of modern speech and audio compression.

The wave model has been used as a principle by industry and manufacturers of gramophone sound systems over many years, where mechanical devices were used to store the wave functions and reproduce the pressure wave. The pressure wave signals are then converted to electronic voltage not only to the speakers but are also stored on various media and transferred via various types of channel. Without the vocoder model, technologies such as Internet telephony and live broadcasts over the World Wide Web today would be impossible.

9.2 Sampling

Digitised audio data are the digital form representation of sounds. They are created by a process of *sampling* followed by *quantisation* (Section 9.3) on the analogue sound wave.

Sampling is a process in which sample values of a continuous signal are taken at a sequence of certain time spans. In other words, sampling is a way of taking certain values at n discrete time $t_1, t_2, \cdots, t_n$. The number of samples taken per time unit is called the *sample rate*.

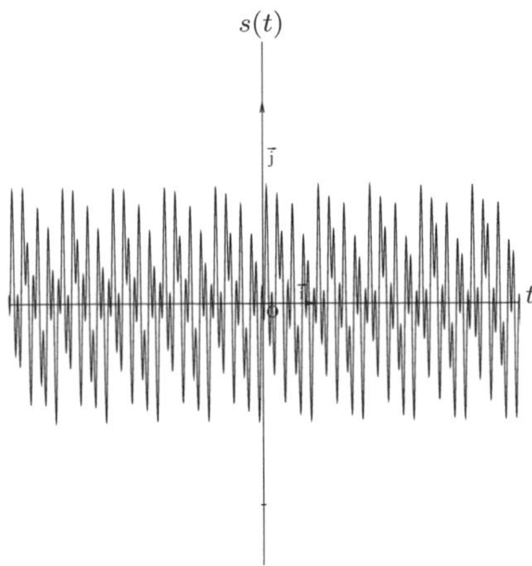

Figure 9.3: Plot $s(t) = 0.1\sin(2\pi t) + 0.3\sin(8\pi t) - 0.25\sin(20\pi t)$

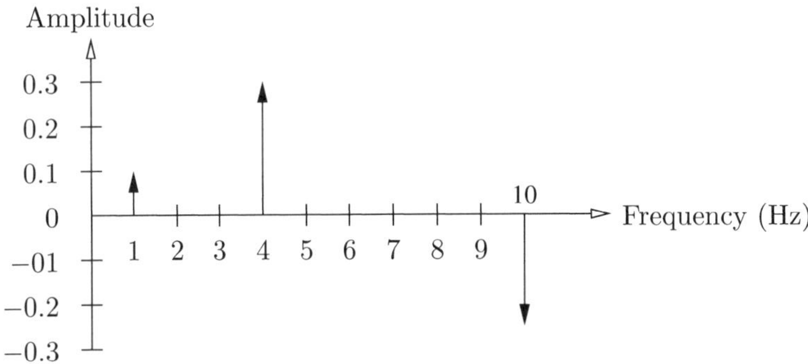

Figure 9.4: Frequency spectrum diagram

Samples may be taken with an equal time interval ($t_i - t_{i-1}$ is a constant for all $i = 1, \cdots, n$) or different time intervals ($t_i - t_{i-1}$ is a variable for each i). Hence there are samples at a *fixed rate*, or samples at a *variable rate* respectively. We assume a fixed sample rate if not otherwise specified in this book.

Example 9.3 *The sampling rate used for audio CDs is 44.1 kHz; 48 kHz is used for DAT (digital audio tape). 22.05 kHz is commonly used for audio signal for delivery over the Internet, and 11.025 kHz may be used for speech.*

Figure 9.5 shows how a set of discrete data $(0.8415, 0.9093, 0.1411, -0.7568, -0.9589, -0.2794, 0, 6570, 0.9894, 0.4121, -0.5440)$ can be obtained by sampling a standard sine signal at $t = 1, 2, 3, 4, 5, 6, 7, 8, 9, 10$.

Note the choice of a right sample rate is critical to the reconstruction of the original signal. Figures 9.6 and 9.7 show the samples at rate 10 and 100 respectively. As we can see, while it is easy to recognise the original signal from Figure 9.7, it is difficult to do so from Figure 9.6. This suggests that the sample

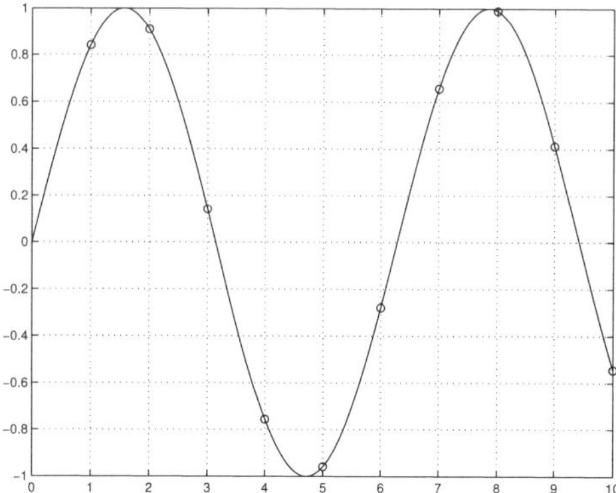

Figure 9.5: Samples at $t = 1, 2, 3, 4, 5, 6, 7, 8, 9, 10$

rate must not be too low.

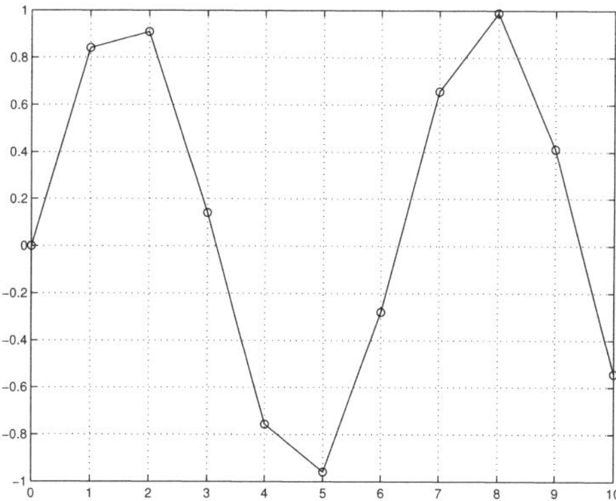

Figure 9.6: Samples at $t = 1, \cdots, 10$

As far as data compression is concerned, we are interested in the minimum number of samples to take in order to reconstruct the original signal. The Nyquist theorem tells exactly how often the samples have to be taken in order to reconstruct the original signal.

9.2.1 Nyquist frequency

The choice of sample frequency rate can directly affect the quality of the reconstructed digital sound.

According to Nyquist theory, if a continuous wave contains a maximum frequency f then the wave must be sampled at a frequency of at least $2f$, twice

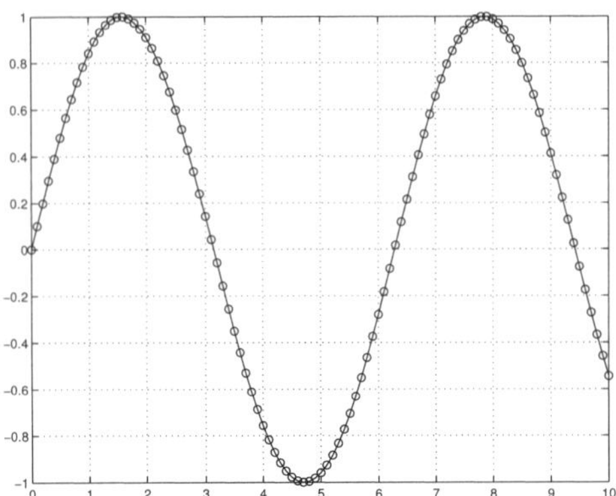

Figure 9.7: Samples at $t = 1, \cdots, 100$

as much as the maximum frequency in order to be able to reproduce the original waveform.

Example 9.4 *Since the human voice normally contains frequencies 0–4000 Hz, the sampling rate should not be lower than* $2 \times 4000 = 8000$ *samples/second.*

Example 9.5 *A communication channel allows signals of frequencies between 1000 Hz to 11 000 Hz. Suppose a signal s(t) can use the channel without any problem. We then know it would be safe to sample the signal using the sample rate* $2 \times 11\,000 = 22\,000$ *samples/second.*

This example can also explain our daily experience to observe a turning wheel with a turning speed under your control. When the wheel starts to turn, say, clockwise slowly, we can tell that the wheel is indeed moving forward clockwise. However, as the wheel turns faster and faster, the wheel appears to turn backward in an anti-clockwise direction. This is because our observation frequency (sampling rate) becomes less than the Nyquist frequency (i.e. twice the frequency of the fast turning wheel in this case) when the wheel turns too fast. We can no longer reconstruct the signal correctly from the observation. Therefore, despite the wheel still turning forward, we get the impression that the wheel is turning backwards.

9.3 Quantisation

The amplitude values obtained after sampling may be long real numbers which are usually rounded to the nearest predefined discrete values. This process of converting the real numbers to the predefined discrete numbers is called *quantisation.*

Figure 9.8 shows an example of the quantisation on the sample data from Figure 9.6 by simply taking $round(t)$ of each sample.

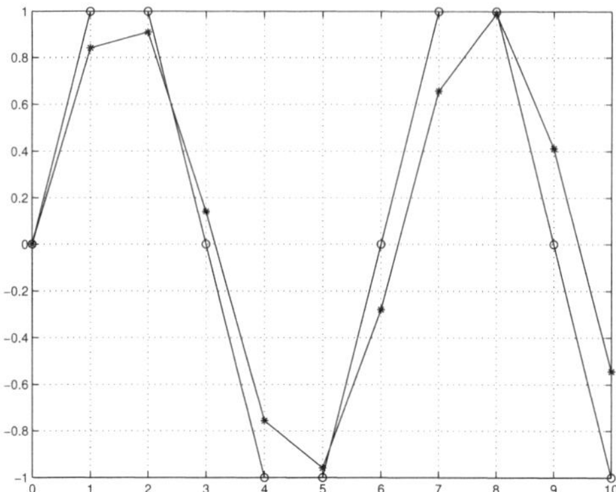

Figure 9.8: Quantisation samples from Figure 9.6

Quantisation does not have to apply to continuous real numbers. It is frequently used in daily life as a means of reducing the number of possible values of any quantity.

Example 9.6 *Students' average marks over all degree modules are usually real numbers. These average marks are usually rounded in order to work out degree classes such as (1st, 2a, 2b, 3rd, pass, fail).*

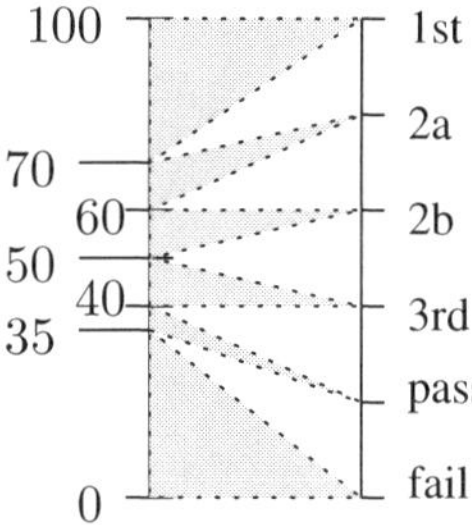

Figure 9.9: Average marks and degree classes

$$m = \begin{cases} \textit{1st} & \textit{if } 70 \leq x \leq 100, \\ \textit{2a} & \textit{if } 60 \leq x < 70, \\ \textit{2b} & \textit{if } 50 \leq x < 60, \\ \textit{3rd} & \textit{if } 40 \leq x < 50, \\ \textit{pass} & \textit{if } 35 \leq x < 40, \\ \textit{fail} & \textit{otherwise.} \end{cases}$$

Example 9.7 *Given a sequence of the real numbers from some sampling (1.1, 2, 3.3, 6.78, 5.48, 4, 3.333, 2.2, 2, 3, 2.1) and predefined integers within a range* *of [0, 10], these sampled real numbers can easily be rounded to (1, 2, 3, 7, 5, 4, 3, 2, 2) by the following simple formula:*

$$n = \begin{cases} x & \textit{if } trunc(x) = trunc(x+0.5), \\ trunc(x)+1 & \textit{otherwise}, \end{cases}$$

where x is a real number, n is the rounded number, and trunc(x) will return the integer part of x.

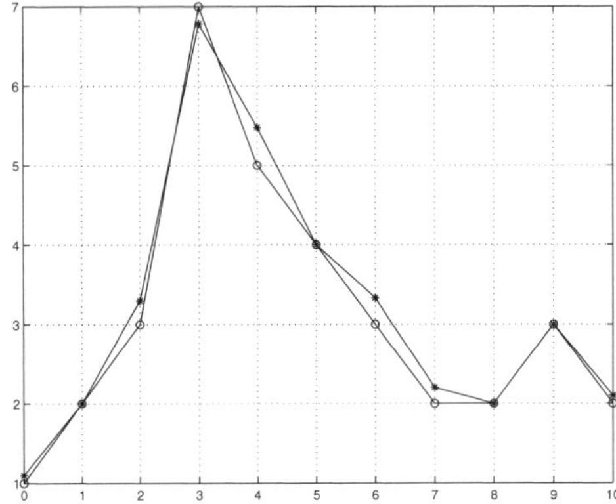

Figure 9.10: Quantisation on (1.1, 2, 3.3, 6.78, 5.48, 4, 3.333, 2.2, 2, 3, 2.1)

As we can see from these examples, certain information is lost during the process of simple quantisation. Given the discrete sequence of data, the original continuous wave function can be reconstructed approximately. Therefore, quantisation is an effective preprocessing for lossy data compression.

The samples can be quantised individually or as a group. A quantisation is called *scalar quantisation* if each of the samples is quantised separately. It is called *vector quantisation* if at least two samples are quantised at the same time.

9.3.1 Scalar quantisation

The amplitude values can be both positive and negative. The function used to map the input sequence of values to the output values is called *quantiser.*

If we plot the output values against the input values of a quantiser, we normally get a staircase shape of curves as shown in Figure 9.11.

There are two types of scalar quantisers. One is called *midrise quantiser* (Figure 9.11(a)) which does not have a zero output level. The other is the so-called *midtread quantiser* (Figure 9.11(b)) where zero is one of the output values.

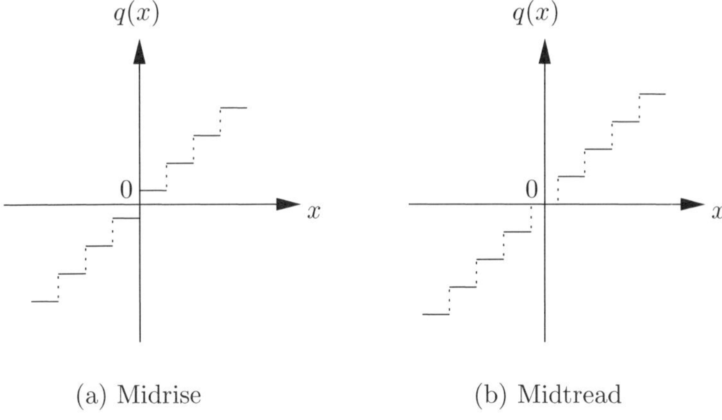

Figure 9.11: Two types of scalar quantisers

9.3.2 Uniform quantisers

A scalar quantiser can be divided into *uniform quantiser* and *non-uniform quantiser*.

Uniform quantisers are those for which the same size of an increment (called *step size*) is used for both its step values for the input x and output $q(x)$. In other words, if k_i and k_{i-1} are two adjacent range values on the x axis, then $q(k_i)-q(k_{i-1}) = k_i-k_{i-1}$ for all $i = 1, 2, \cdots$ (Figure 9.12(a)). The reconstructed values are usually the midpoint of the two adjacent step values $(k_i - k_{i-1})/2$ and $(q(k_i) - q(k_{i-1}))/2$ (Figure 9.12(b)).

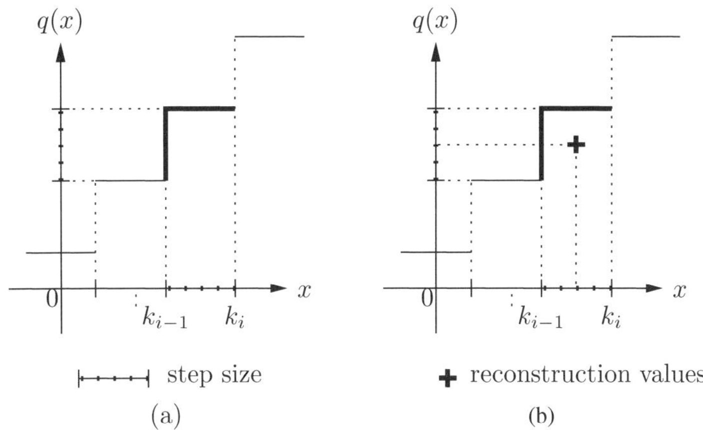

Figure 9.12: Uniform scalar quantiser

Suppose the reconstruction values are r_i, where $i = 1, 2, \cdots$. The difference between quantised and unquantised values is called *quantisation error*, or *round-off error*. The error e_q can be expressed as $e_q = \hat{x} - x$.

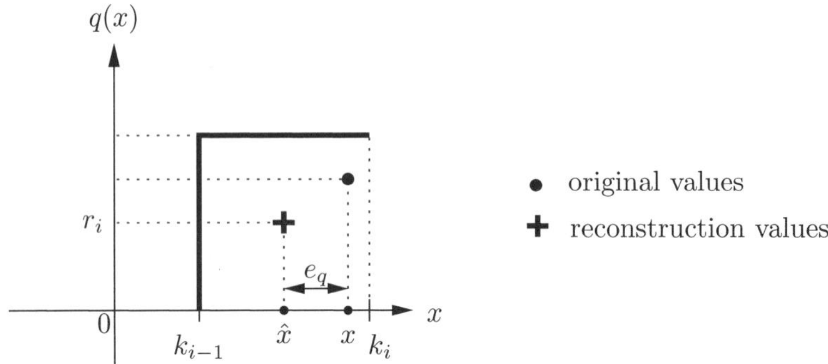

Figure 9.13: Quantisation error

Figure 9.13 shows a quantisation error e_q between the original sample value and the value produced by the quantiser. This indicates the distortion caused by the quantiser. The most common distortion measure is the squared error distortion for each error:

$$d(x, \hat{x}) = e_q^2 = (\hat{x} - x)^2$$

The mean squared error distortion is used to measure the *average distortion* for the quantiser:

$$\sigma^2 = \frac{1}{N}\sum_{i=1}^{N} e_q^2 = \frac{1}{N}\sum_{i=1}^{N}(\hat{x}_i - x_i)^2$$

9.3.3 Non-uniform quantisers

Non-uniform quantisers use *variable* step sizes. A non-uniform quantiser applies different step sizes to different amplitudes of an input signal.

One way to achieve this is to use a *look-up* table for the step sizes. Another way is to use a non-linear monotonically increasing function to define the input and output of the quantiser.

For example, two common functional types are used in non-uniform quantisers: *power-law companding* and *logarithmic companding*. The word *companding* is derived from the words *com*pressing and ex*panding* to reflect the two activities involved. The idea is to model the situation where a small change on the low input value and a large change on the high input value can lead to similar sized steps on the output of the quantiser. Companding techniques reduce the noise and crosstalk level at the sound receiver.

In power-law companding, a power function is used as below:

$$c_{power}(|x|) = |x|^p$$

In logarithmic companding, a logarithmic function $\log(x)$ is used.

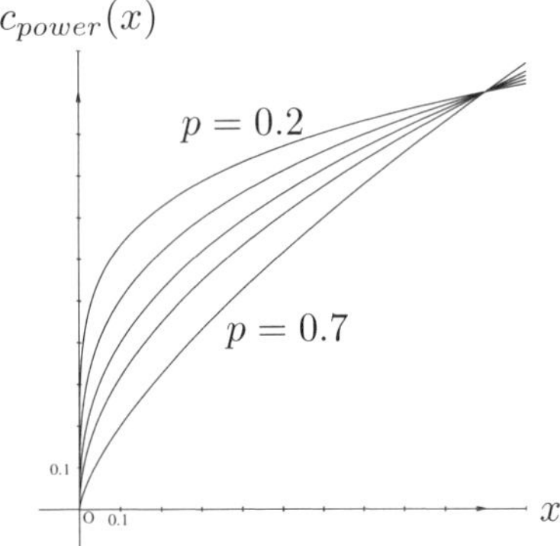

Figure 9.14: Power law companding functions when $p = 0.2, 0.3, 0.4, 0.5, 0.7$

A common form of logarithmic companding is the so-called μ *law*

$$c_\mu(|x|) = \frac{\log_b(1 + \mu|x|)}{\log_b(1 + \mu)}$$

and another common form is the so-called A *law*:

$$c_A|x| = \begin{cases} \frac{1+\ln(A|x|)}{1+\ln(A)} & \text{for } |x| > 1/A \ , \\ \frac{A}{1+\ln(A)}|x| & \text{for } |x| \leq 1/A \ . \end{cases}$$

These last two laws, i.e. μ law and A law companding, have many applications in the area of telecommunication and are included in the CCITT (previously Telephone and Telegraph Consultative Committee, now known as the ITU-T, International Telecommunication Union Telecommunications Sector). The particular values of $\mu = 255$ and $A = 87.56$ are used in the standard.

As we see from the previous discussion, the audio compression process starts from sampling the analogue audio signal. Similarly, the decompression process does not complete until the audio analogue signal in wave form strikes the human ear. Every stage is important to the performance of compression and decompression but it is the first and last stages which are the most effective and important.

Figure 9.15 shows a block diagram of a typical audio coding system.

It is impossible to cover all these interesting techniques in this book. Nevertheless, we give a flavour of some of the basic approaches.

9.4 Compression performance

For lossless compression, all we need to measure the compression performance is the compression ratio. With lossy compression, we have to balance the compression ratio and the quality of the reconstructed sound. Since the sound is a function of time, we also have to take the time into account in addition to

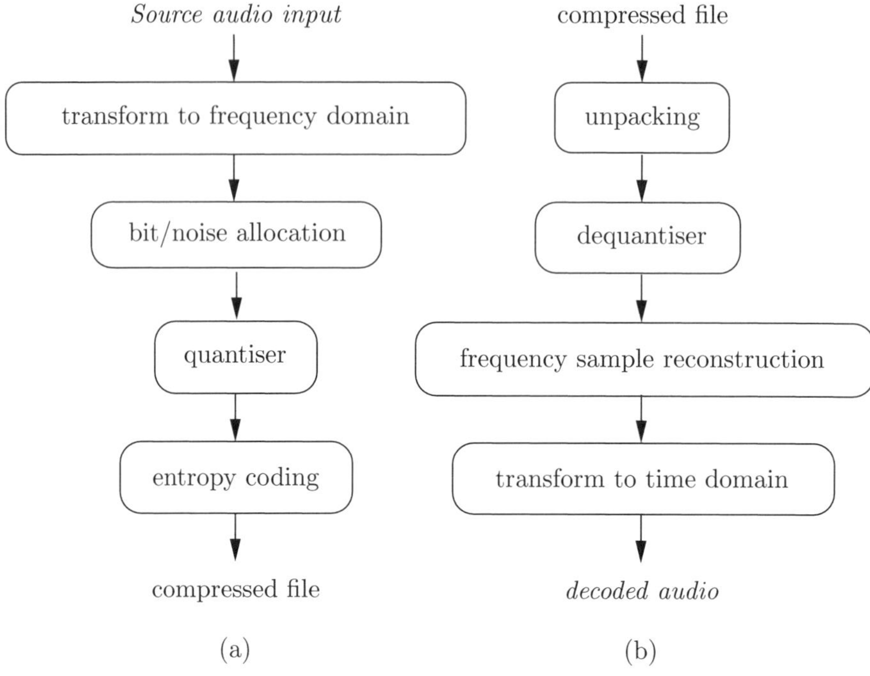

Figure 9.15: Audio encoder (a) and decoder (b)

the normal complexity consideration. Over the years, the most important factors to be considered in audio compression include *fidelity*, *data rate*, *complexity* and *delay*. The balance among these factors normally varies depending on the application being supported.

1. Fidelity measures how close perceptually the reconstructed audio signal sounds in comparison to the original signal.

2. Data rate represents the speed of data transmission via a communication channel, usually measured in bits/second. This measure is required due to the restriction of various media such as the access speed to various storage, the capacity of a transmission channel and the playback speed of a certain mechanical device.

3. Complexity means the amount of work required and consequently its cost in order to achieve a certain compression or decompression task. The cost is not always reflected in the amount of work because of the increasing computer power and the development of technology. In the real world, perhaps the implementation cost is more important than anything else.

4. Delay is critical in a real-time application such as telephony, or teleconferencing.

These measures may be quite subjective but this is due to the nature of audio systems. The audio compression methods rely a lot upon the perspectiveness of the human audio system. The research in this area relies on multidisciplinary knowledge and techniques much more heavily than any other areas. As in any other complex system design, any audio compression system design in general aims at high fidelity with low data rates, while keeping the complexity and delay as low as possible.

Two types of sound with distinct characteristics are *speech* and *music*. They are the most commonly used types of sound in multimedia productions today. For example, if we plot the amplitude over a period of time for both the sound generated by speech and by a piece of music, we would find that the shape of the two waves is quite different.

The requirements to media by the two types of audio data are also different. For example, the media for *telephone speech* needs to be able to handle signals with a frequency of 200–3400 Hz, and *wideband audio* with frequency 50–7000 Hz, while the media for music corresponds to the *CD-quality audio* needed to be able to process the signals with frequency 20–20 000 Hz.

Representations specific to speech and music have been developed to effectively suit their unique characteristics. For example, speech may be represented in a sound model that is based on the characteristics of the human vocal apparatus, but music can be represented as instructions for playing on virtual instruments.

This leads to two big compression areas, namely *speech compression* and *music compression*. They have been developed independently for some time. Conventionally, voice compression aims at removing the silence and music compression at finding an efficient way to reconstruct music to play to the end user. Today, almost every stage between the source sound and the reconstructed sound involves a data compression process of one type or another.

9.5 Speech compression

This is also called *voice compression*. Research on speech compression started to produce amazing results as early as 1928 by Homer Dudley, an engineer at Bell Laboratories. His idea was to compress a speech signal with a bandwidth of over 3000 Hz into the 100 Hz bandwidth of a new transatlantic telegraph cable. Instead of sending the speech signal itself, he sent a specification of the signal to the receiver.

We only briefly introduce a few commonly used compression methods here.

9.5.1 Speech coders

Here two major types of audio data are considered:

- telephone speech
- wideband speech.

The goal is to achieve a compression ratio of 2:1 or better for both types of data. More specifically, the aim is to compress telephone speech to the compressed bit rate of less than 32 kbps, and to 64 kbps for wideband speech.

The following constraints make it possible to fulfill some of the tasks:

- The human ear can only hear certain sounds in normal speech.
- The sounds produced in normal human speech are limited.
- Non-speech signals on telephone lines are noises.

Two coders are used in practice, namely *waveform coders* and *vocoders* (voice coders). They both apply detailed models of the human voice tract to identify certain types of patterns in human speech.

9.5.2 Predictive approaches

The idea is to try to predict the next sample based on the previous sample and code the differences between the predicted and the actual sample values.

A basic version of the implementation would simply apply entropy coding, such as Huffman or arithmetic, to the differences between successive samples.

This approach is used in the so-called VOCPACK algorithm for compressing 8 bit audio files (.wav).

Example 9.8 *Given a series of sample data, (27, 29, 28, 28, 26, 27, 28, 28, 26, 25, 27), we can write the difference* $(2, -1, 0, -2, 1, 1, 0, -2, -1, 2)$*. We then derive the alphabet* $(2, 1, 0, -1, -2)$ *with the frequencies (2, 2, 2, 2, 2) respectively. The entropy of the distribution of differences is, therefore,* $5 \times 0.2 \log_2 5 = 2.32$ *bits. Compared to a 8 bit coding scheme, this would give a compression ratio of 8:2.32, i.e. 3.45:1 if applied to the entire file.*

Of course, sampling is also a critical step to achieving good compression and it is always worth considering.

9.5.3 Silence compression

This approach applies the run-length to compress the silence in sound files. However, the whole process is a *lossy* approach since quantisers or filters may be used to preprocess the relative silence and noise in the sound file. This includes finding a suitable

1. threshold value for defining silence
2. coder for silence
3. coder for the start and end of the silence.

9.5.4 Pulse code modulation (ADPCM)

This technique is well known in other research areas such as telecommunication and networking. ADPCM stands for Adaptive Differential Pulse Code Modulation. It was formalised as a standard for speech compression in 1984 by ITU-T. The standard specifies compression of 8 bit sound (typical representation for speech) sampled at 8 kHz to achieve a compression ratio of 2:1.

9.6 Music compression

Compression algorithms for music data are less well developed than those for other types of data.

The following facts contribute to this situation:

- There are various formats for digital sounds on computer.
- The development of digital audio has taken place in recording and broadcast industries.
- There are currently three major sound file formats: AIFF for MacOS, WAV or WAVE for Windows and AU (NeXT/Sun audio file format). Each of the three has evolved over the years and now provides similar capabilities in terms of sampling rates, sizes and CD and DAT standard value storage, while MP3 in its own file format cannot accommodate sound compression by any other method.

We usually use a high bit rate and often attempt to capture a wider range of frequencies (20 to 20 kHz). It is usually difficult to decide what is a good compression system because this depends on each person's hearing ability. For example, to most people, lower frequencies add bass resonance and the result would sound more like the human voice. This, however, may not be the case at all for others.

9.6.1 Streaming audio

The idea of streaming audio is to deliver sound over a network and play it as it arrives without having to be stored on the user's computer first. This approach is more successful in general for sound than it is for video due to the lower requirement for bandwidth.

The available software includes:

- Real Networks' RealAudio (companion to RealVideo)
- Streaming QuickTime
- 'lo-fi' MP3.

These are used already for broadcasting live concerts, for the Internet equivalent of radio stations, and for providing a way of playing music on the Internet.

9.6.2 MIDI

The main idea of MIDI compression is, instead of sending a sound file, to send a set of instructions for producing the sound. Of course, we need to make sufficiently good assumptions about the abilities of the receiver to make sure the receiver can play back the sound following the instruction, otherwise the idea cannot work in practice.

MIDI stands for Musical Instruments Digital Interface. It was originally a standard protocol for communicating between electronic instruments. It allowed instruments to be controlled automatically by sound devices that could be programmed to send out MIDI instructions.

MIDI can control many sorts of instruments, such as synthesisers and samplers, to produce various sounds of real musical instruments.

Examples of available software include:

- QuickTime
- Cakewalk Metro
- Cubase

QuickTime incorporates MIDI-like functionality. It has a set of instrument samples and can incorporate a superset of the features of MIDI. It can also read standard MIDI files, so any computer with it installed can play MIDI music without any extra requirement.

Note: one should realise that sound tends to work together with pictures or animation. In future most audio work has to take any potential synchronisation into consideration. For example, sound divided into video frames by some time code would be a useful function for film editing.

Summary

Digital audio compression offers good motivation for lossy techniques such as sampling and quantisation. Sound can be viewed and modelled in a collection of sine and cosine waves. The compression techniques covered so far can be applied at various compression stages and specific situations. Techniques are centred at two application areas: speech compression and music compression.

Learning outcomes

On completion of your studies in this chapter, you should be able to:

- explain how sound can be represented by a periodic function
- illustrate sound by a frequency spectrum diagram as well as the normal plot for periodic functions
- describe the concept or principle of terms in audio data compression

- outline the distinction between voice compression and music compression in terms of the issues concerned in audio compression
- describe the main ideas of MIDI compression.

Exercises

E9.1 Given that singing has characteristics of both speech and music, which compression algorithms would you expect to be most successful on songs?

E9.2 Given a frequency spectrum diagram as in Figure 9.16, write the signal $s(t)$ in its analogue form, i.e. represented by the sum of sine and cosine waves.

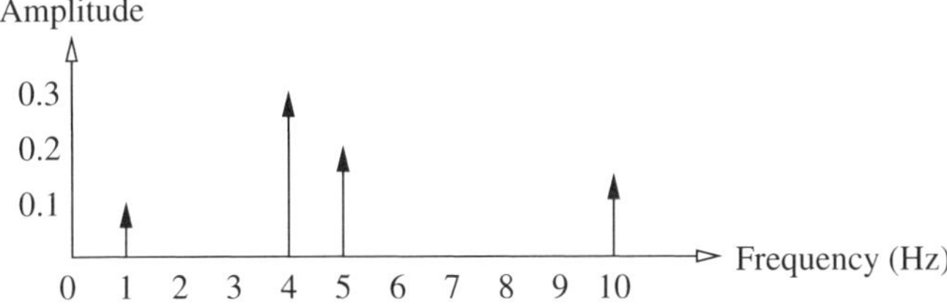

Figure 9.16: A frequency spectrum diagram

E9.3 Sketch the following discrete signals: :

(a) $x(n) = -5, -4, -3, -2, -1, 1, 2, 3, 4$, for $n = 0, 1, \cdots, 7$

(b) $y(n) = 0, -1, -2, 0, 1, 2, 0, -1, -2$, for $n = -3, -2 \cdots, 0, \cdots 4$

(c) $2x(n)$, $x(n-3)$, $2y(n)$, $-3x(n-2) + 2y(n+1)$

(d) $x(t) = 3sin(2\pi t/8)$, $-8 < t < 8$

(e)

$$x(n) = \begin{cases} 1 & \text{if } n \leq -3, \\ 0 & \text{if } 0 < n \leq 3, \\ -3 & otherwise \end{cases}$$

E9.4 A signal can be decomposed to three basic sine waves, $\sin(2\pi t)$, $\sin(2.5t)$ and $\sin(5\pi t)$. What sampling rate should be used according to the Nyquist theorem?

E9.5 Explain what MIDI stands for. Write a short essay of about 500-1000 words to introduce any application software which uses MIDI in one way or another.

Assessment

S9.1 Explain briefly the following terms:

(a) Sampling
(b) ADPCM
(c) MIDI.

S9.2 Explain how a frequency spectrum diagram can be used to represent function $f(t) = 0.5 + \sin(880\pi t) + \sin(1760\pi t)$.

Bibliography

[BG02] M. Bosi and R.E. Goldberg. *Introduction to Digital Audio Coding and Standards.* Kluwer Academic Publisher, 2002.

[Ger77] A. Gersho. Quantization. *IEEE Communications Magazine*, 15:16–29, September 1977.

[GG91] A. Gersho and R.M. Gray. *Vector Quantization and Signal Compression.* Kluwer Academic Publishers, Norwell, MA, 1991.

[Gra84] R.M. Gray. Vector quantization. *IEEE Transactions on Acoustics, Speech and Signal Processing*, 1:4–29, April 1984.

[Jay73] N.S. Jayant. Adaptive quantization with one word memory. *Bell System Technical Journal*, 52:1119–1144, September 1973.

[Jay76] N.S. Jayant. *Waveform Quantization and Coding (ed).* IEEE Press, New York, 1976.

[JN84] N.S. Jayant and P. Noll. *Digital Coding of Waveforms: Principles and Applications in Speech and Video.* Prentice Hall, Englewood Cliffs, New Jersey, 1984.

[MRG85] J. Makhoul, S. Roucos, and H. Gish. Vector quantization in speech coding. *Proceedings of the IEEE*, 73:1551–1588, 1985.

[Sch99] M. R. Schroeder. *Computer Speech: Recognition, Compression, Synthesis.* Springer, 1999.

[Swa] P.F. Swaszek. Vector quantization. In *I.F. Blake and H.V. Poor (eds), Communications and Networks: A Survey of Recent Advances*, pages 362–389, Springer-Verlag, New York (1986).

[Tre78] H.M. Tremaine. *Audio Cyclopedia.* 7th edition, 1978.

Chapter 10

Image compression

In this chapter, we introduce compression methods for a different type of data, namely, digital images or images for short. General compression methods may be applied to images. However, certain characteristics of image data, if identified, often lead to more effective data compression algorithms. We first consider the *image data*.

10.1 Image data

By image data we actually mean the representation of real-life graphics in digital form which can be processed by conventional computers. The word *digital* here means that the data in this representation are discrete. For example, the display area of a conventional computer screen consists of a large number of small discrete units called *pixels*. The pixels are mapped onto a two-dimensional array of data entries of a certain type. Each entry is represented in the form of 0s and 1s.

We often adopt the term *pixels* to mean these binary data. The two-dimensional arrays are also called 'colour maps', for they are usually used to control certain colour display systems such as computer screens or printers. Figure 10.1 shows a tiny proportion of an image including 12×8 pixels of a circle image.

Two characteristics of image data are of quantity and quality. The first characteristic of images is the massive amount of data involved in almost every application. Images are stored in files and they tend to be much bigger in size compared to text files. For text files in ASCII code, a book of a million words may occupy about 5 million bytes, that is 5 MB. In contrast, one image file can easily be a thousand to a million times bigger.

The second characteristic of images is that the quality of an image depends not only on the image data but also on the display device and the sensation of the human visual system.

Most people these days have the experience of using some painting programs.

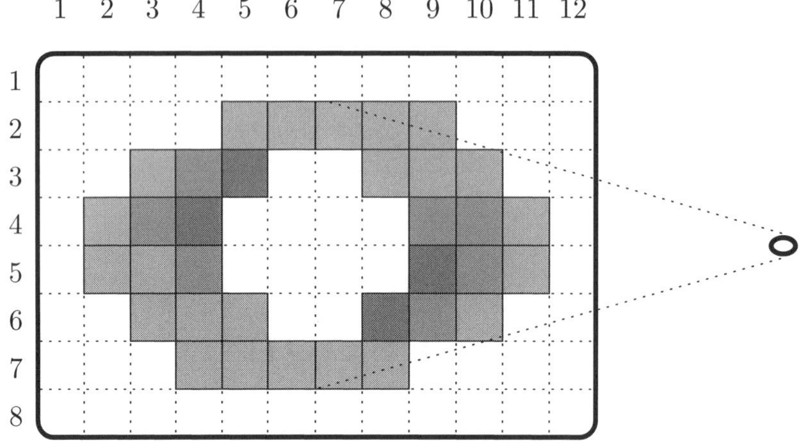

Figure 10.1: Image pixels

For example, Figure 10.2 shows a small circle which was painted using the program KolourPaint.[1] Figure 10.3 shows the same paint but it was enlarged by 800% of the original. As we can see from the latter, the circle consists of many small squares of either white or grey coloured pixels.

10.1.1 Resolution

Every display device requires a certain number of pixels to be able to show something easily recognisable by the human visual system. On the other hand, each device is restricted by the number of pixels that it can actually handle. The maximum number of pixels for a display is called the *resolution* of the display device. Often the higher the resolution for a display on a limited display area, the better the image display quality can be achieved. A digital image can also be measured in terms of the *resolution*, which measures how finely a device approximates continuous images using a finite number of pixels.

Let us look at two common ways of describing resolution: one is so-called *dpi* and the other *pixel.*

The term *dpi* stands for *dots per inch.* It represents the number of dots per unit length for data from devices such as printers or scanners. The term *pixel dimension* measures the number of pixels per frame for the video data from digital cameras.

For TV systems, the PAL frame is 768 by 576 pixels and the NTSC frame is 640 by 480. The most common standard resolutions on a conventional PC monitor are 1024×768 (786 432 pixels), 1024×960, 1280×1024, 1400×1050, and so on. Some old monitors can only support a resolution of 640×480 (pixels).

[1] The screen shot was done using a program called KSnapshot under Linux.

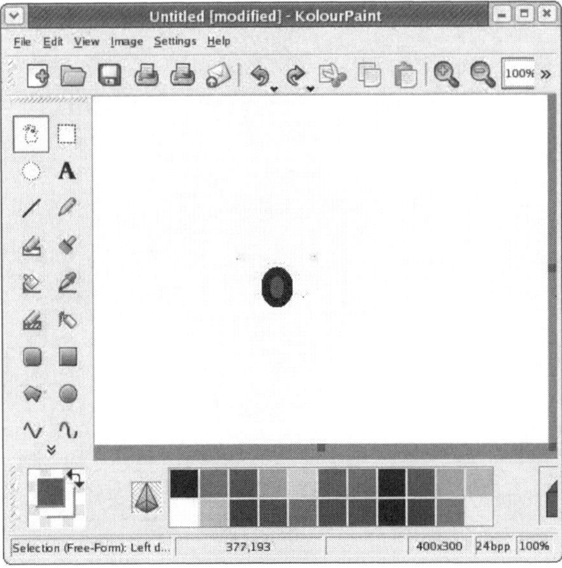

Figure 10.2: Using a paint program

For each frame of image to display, the colour information related to each pixel is stored in a temporary memory storage area called a *frame buffer*. Each pixel on a display device is mapped to a memory cell. A control program takes care of the routine signals to the display device and the data in the frame buffer. The control information about the display device is also stored and can be accessed by an application program. Often the monitor and video adaptor restrict the number of pixels of a computer system.

In the multimedia world today, two types of image are most commonly used: one is the so-called *bitmap image* and the other is *vector graphics*.

10.2 Bitmap images

These are also called *photographic* images for two-dimensional pictures. A bitmap image is an array of pixel values. The data are the values of the pixels. Many bitmap images are created from digital devices such as scanners or digital cameras, or from programs such as Painter which allows visual artists to paint images.

10.2.1 Displaying bitmap images

Since a bitmap image is in fact represented in a computer by an array of pixel values, there has to be a way of mapping each pixel value to the physical dots on a display screen.

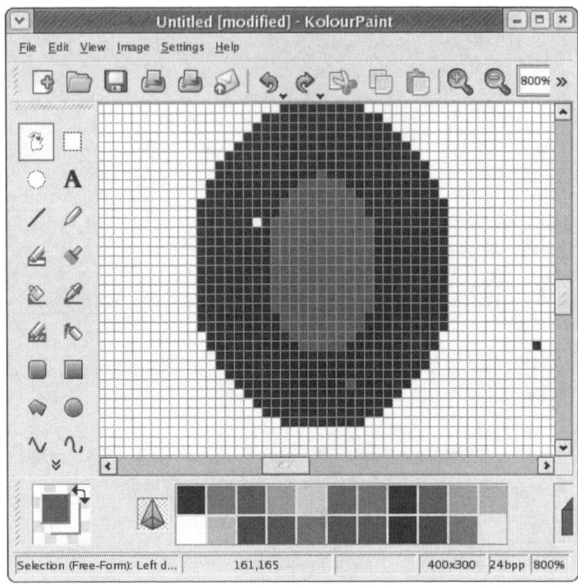

Figure 10.3: An enlarged image

In the simplest case, each pixel corresponds one-to-one to the dots on the screen. In the general case, the physical size of the display of an image will depend on the device resolution. The pixel values are stored at different resolutions from the displayed image. Computations for scaling and clipping usually need to be performed to display a bitmap image.

In general, we have the following formula to decide the image size and scaling:

```
imageDimension = pixelDimension/deviceResolution,
```

where the `deviceResolution` is measured in *dpi*.

10.3 Vector graphics

The image in vector graphics is stored as a mathematical description of a collection of graphic components such as lines, curves, shapes and so on.

Vector graphics are normally more compact, scalable, resolution independent and easy to edit. They are very good for three-dimensional models on the computer, usually built-up using shapes or geometric objects that can easily be described mathematically.

10.3.1 Storing graphic components

The graphic components are stored by their mathematical functions instead of pixels. For example, consider storing a line on the computer.

Example 10.1 *The mathematical function for a straight line is $y = kx + b$, where k is the slope and b the intercept. The function, instead of all the pixels along the line, can be stored for the line.*

Since a geometric line is actually a finite segment of a line, we also need to store the two end points of the segment. In fact, it is sufficient only to store the coordinates of the two end points because we know the line function can be represented by two point coordinates. That is, given (x_1, y_1) and (x_2, y_2), we can easily write the function for the line: $y = (y_2 - y_1)/(x_2 - x_1)x + (x_2y_1 - x_1y_2)/(x_2 - x_1)$.

10.3.2 Displaying vector graphic images

This would require some computation to be performed in order to interpret the data and generate an array of pixels to be displayed. The process of generating a pattern of pixels from a model is called *rendering*.

If the image is a line, only two end points are stored. When the model is rendered for display, a pattern of pixels has to be generated to display the line with the two end points.

10.4 Bitmap and vector graphics

The differences between bitmap images and vector graphics can be very obvious in terms of visual characteristics, but we are concerned more about the following issues:

1. The **requirements** of the computer system: a bitmap image must record the value of every pixel, but vector description may take much less space for an image with simple structure, so may be more economical.
2. The **size** of a bitmap image file depends on the display resolution. It is independent of the complexity of the image.

   In contrast, the size of vector graphics depends on the number of objects of which the image consists. It is independent of any resolution.
3. The **approach** of the so-called *painting* programs produces bitmap images while that of *drawing* programs produces vector graphics.
4. The **behaviour** of both bitmap images and vector graphics is different when resized or scaled.

Most graphic applications today require a combination of bitmap and vector graphics. A transformation between vector graphics and bitmap images may be necessary. The following two processes are usually implemented for the transformation:

Rasterising

A raster is a predetermined pattern of scanning lines to provide substantially uniform coverage of a display area. Rasterising is the process of interpreting the vector description of graphics for a display area. A vector graphic loses all the vector properties during the process, e.g. the individual shapes can no longer be selected or moved because they become pixels.

Vectorisation

This is a more complicated process of transforming pixels to vectors. The difficulties arise from the need to identify the boundary of a pixel image using the available data of curves and lines and to colour them. The vector file tends to be much bigger than the pixels file before vectorisation.

10.5 Colour

As we know, colour is a subjective sensation experience of the human visual system and the brain.

10.5.1 RGB colour model

This is a colour representation model for computing purposes. It comes from the idea of constructing a colour out of so-called *additive primary colours* (i.e. Red, Green and Blue or the RGB for short). Although there is no universally accepted standard for RGB, the television and video industries do have a standard version of RGB colour derived from Recommendation ITU-R BT.709 for High Definition TV (HDTV). Monitors have been built increasingly to follow the recommendation.

In the RGB model, we assume that all colours can in principle be represented as combinations of certain amounts of the red, green and blue. Here the amount means the proportion of some standard of primary red, green and blue.

Example 10.2 *Suppose that the proportion is represented as percentages. (100%, 0%, 0%) then represents a colour of 'pure' red, and others*

- *(50%, 0%, 0%) a 'darker' red*
- *(0%, 0%, 100%) a 'pure' blue*
- *(0%, 0%, 0%) black*
- *(100%, 100%, 100%) white*

  ⋮

  and so on.

In fact, there are two commonly used representations of digital colours: they are so-called *RGB representation* and *LC representation.*

10.5.2 RGB representation and colour depth

Since it is only the relative values of R, G and B that matter, we can actually choose any convenient value range, as long as the range provided sufficiently distinguishable values.

In a common RGB representation, we use 1 byte (8 bits) to represent the brightness of each of the three primary colours, which gives a range of [0, 255] (or [1, 256]). Three bytes ($8 \times 3 = 24$ bits) describe each pixel of a colour image. In this way, 256^3 (i.e. 16 777 216) different colours can be represented.

The number of bits used to hold a colour value is often called *colour depth.* For example, if we use 24 bits to represent one pixel, then the colour depth is 24. We sometimes also say '24 bit colour' to refer to a colour with 24 bit colour depth.

Example 10.3 *Some colours from the Linux colour database are 24 bit colours:*

R	*G*	*B*	*colour*	*R*	*G*	*B*	*colour*
255	*255*	*255*	*white*	*255*	*250*	*250*	*snow*
0	*0*	*128*	*navy blue*	*248*	*248*	*255*	*ghost white*
0	*0*	*255*	*blue1*	*255*	*239*	*213*	*papaya whip*
0	*255*	*0*	*green*	*255*	*228*	*225*	*misty rose*
...							

Colour depth determines the size of the bitmap image: each pixel requires 24 bits for 24 bit colour, but just a single bit for 1 bit colour. Hence, if the colour depth is reduced from 24 to 8, the size of a bitmap image will decrease by a factor of 3 (ignoring any fixed-size housekeeping information).

10.5.3 LC representation

This is another common representation which is based on *luminance* (Y) and *chrominance* (C) values. Luminance Y reflects the brightness, and by itself gives a greyscale version of the image.

The approach is based on colour differences. The idea is to separate the brightness information of an image from its colour. By separating brightness and colour, it is possible to transmit a picture in a way that the colour information is undetected by a black and white receiver, which can simply treat the brightness as a monochrome signal.

A formula is used which has been empirically determined for the best greyscale likeness of a colour image.

$$Y = 0.299R + 0.587G + 0.114B$$

The chrominance (colour) components provide the additional information needed to *convert* the greyscale image to a colour image. These components are represented by two values C_b and C_r given by $C_b = B - Y$ and $C_r = R - Y$.

This technique was initially used during the development from black-and-white to colour TV. However, it turns out to have some advantages. For example, since the human eye is much better at distinguishing subtle differences in brightness than subtle differences in colour, we can compress the Y component with greater accuracy (lower compression ratio), and make up for it by compressing the C_b and C_r components with less accuracy (higher compression ratio).

10.6 Classifying images by colour

For compression purposes, we can classify the images for compression by number of colours used. Three approaches are available, namely *bi-level image*, *greyscale image*, and *colour image*.

Bi-level image

This is actually the image with 1 bit colour. A single bit allows us to distinguish two-different colours.

Images are captured by scanners using only *two* intensity levels, one is the 'information' and the other the 'background'. Applications include:

- Text, line drawings or illustrations
- FAX documents for transmission
- Sometimes photographs with shades of grey.

Greyscale image

This is actually the image with an 8 bit colour depth. The images are captured by scanners using multiple intensity levels to record shadings between black and white. We use 8 bits to represent one pixel, and hold one colour value to provide 256 different shades of grey. Greyscale images are appropriate for medical images as well as black-and-white photographs.

Colour image

Colour images are captured by (colour) scanners, using multiple intensity levels and filtering to capture the brightness levels for each of the primary colours, R, G and B (Red, Green, Blue).

Some computer systems use 16 bits to hold colour values. In this case, either 1 bit is left unused or different numbers of bits are assigned to R, G and B. If it is the latter, then we usually assign 5 bits for R and B, but 6 bits for G. This allocation is due to the fact that the human eye is more sensitive to green light than to red and blue.

Most common colour images are of a 24 bit colour depth. Although 24 bits are sufficient to represent more colours than the eye can distinguish, higher colour depths such as 36 or even 48 bits are increasingly used, especially by scanners.

10.7 Classifying images by appearance

Images can be classified by their appearances which are caused by the way in which colours distribute, namely *continuous-tone* image, *discrete-tone* image and *cartoon-like* image.

Continuous-tone image

This type of image is a relatively natural image which may contain areas with colours. The colours seem to vary continuously as the eye moves along the picture area. This is because the image usually has many similar colours or greyscales and the eye cannot easily distinguish the tiny changes of colour when the adjacent pixels differ, say, by one unit.

Examples of this type of image are the photographs taken by digital cameras or the photographs or paintings scanned in by scanners.

Discrete-tone image

Alternative names are *graphical image* or *synthetic image*. This type of image is a relatively artificial image. There is usually no noise and blurring as there is in a natural image. Adjacent pixels in a discrete-tone image are often either identical or vary significantly in value. It is possible for a discrete-tone image to contain many repeated characters or patterns.

Examples of this type of image are photographs of artificial objects, a page of text, a chart and the contents of a computer screen.

Cartoon-like image

This type of image may consist of uniform colour areas but adjacent areas may have very different colours. The uniform colour areas in cartoon-like images are regarded as good features for compression.

Observation

From the above discussion, we can see that:

1. Image files are usually large because an image is two dimensional and can be displayed in so many colours. In a bitmap image, each pixel requires typically 24 bits to represent its colour.
2. The loss of some image features is totally acceptable as long as the human visual system can tolerate. After all, an image exists only for people to view. It is this fact that makes lossy compression possible.

3. Each type of image contains a certain amount of redundancy but the cause of the redundancy varies, and this leads to different compression methods. That is why so many different compression methods have been developed.

10.8 Image compression

Image compression can be lossless or lossy, although most existing image compression systems are lossy. A typical lossy image compression system consists of several components applying several compression approaches.

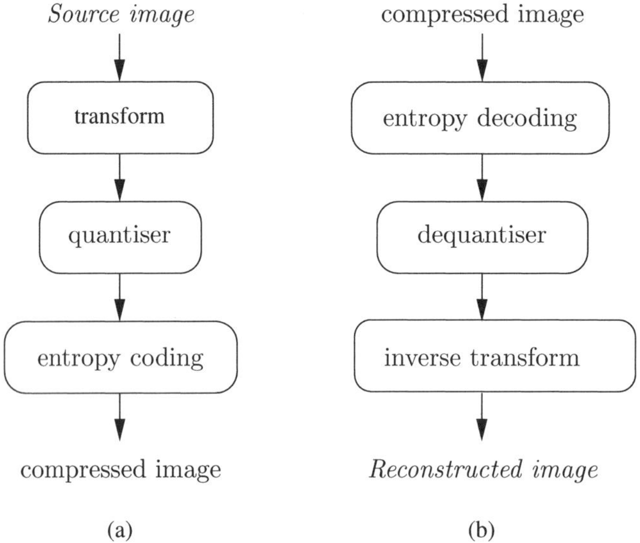

Figure 10.4: A typical image compression system

Figure 10.4(a) shows a typical encoding process and Figure 10.4(b) a typical decoding process. During encoding, the image data are first divided into manageable small units called *blocks*, say 8×8. The transform techniques (Chapter 8) are then applied to the source image blocks. For DCT or DWT, coefficients are actually transformed. This is followed by scalar or vector quantisation to reduce the number of bits required. Finally, an entropy encoder such as run-length, Huffman or arithmetic encoder is applied to achieve an even better overall compression effect.

Two facts determine the main issues in image compression. First, the images in many applications do not need to be reproduced exactly the same as the original due to the tolerance of the human visual system. The information that cannot be perceived or noticed is regarded as *irrelevant* information to the image viewer. Secondly, the neighbouring pixels in images are highly correlated and redundant. The correlation between neighbouring pixel values is called *spatial redundancy*, and the correlation between different colour planes is called *spectral redundancy*. Hence, the fundamental tasks in image compression are to reduce the irrelevancy and correlation in the images.

We shall introduce the general approaches in both classes to the image data discussed in the previous section.

First, we look at lossless image compression.

10.8.1 Lossless image compression

Many techniques for text data can be extended and applied to image data. These lead to many lossless approaches for image compression.

We first look at image compression for a binary source, i.e. bi-level images. Bi-level images are efficient for certain applications where the visual satisfaction is not a priority. For example, data transmission speed may be more important than anything else in telecommunication. The fax standards such as the ITU-T T.4 and T.6 recommendations are still in use today.

For bi-level image data, two popular approaches, *run-length coding* and *extended approach*, are frequently used.

10.8.1.1 Run-length coding

The run-length approach can be applied to bi-level images due to the fact that: each pixel in a bi-level image is represented by 1 bit with two states, say black (B) and white (W), and the immediate neighbours of a pixel tend to be in an identical colour.

Therefore, the image can be scanned row by row and the length of runs of black or white pixels can be computed. The lengths are then encoded by variable-size codes and are written into the compressed file.

Example 10.4 *A character* **A** *represented by black-and-white pixels:*

```
  1234567890123456789012345
1 BBBBBBBBBBBBBBBBBBBBBBBBB
2 BBBBBBBBBBBwwwBBBBBBBBBBB
3 BBBBBBBBBBwwBwwBBBBBBBBBB
4 BBBBBBBBBwwBBBwwBBBBBBBBB
5 BBBBBBBBwwBBBBBwwBBBBBBBB
6 BBBBBBBwwBBBBBBBwwBBBBBBB
7 BBBBBBwwwwwwwwwwwwwBBBBBB
8 BBBBBwwBBBBBBBBBBBwwBBBBB
9 BBBBwwBBBBBBBBBBBBBwwBBBB
0 BBBBBBBBBBBBBBBBBBBBBBBBB
```

Solution We notice that line 4 consists of several *runs* of pixels: 9B, 2w, 3B, 2w, 9B, i.e. 9 'B's followed by 2 'w's, followed by 3 'B's, then 2 'w's and then 9 'B's. Similarly, line 5 consists of 8B, 2w, 5B, 2w, 8B, and so on.

A better compression may be achieved by taking into consideration the *correlation* of adjacent pairs of lines.

For example, a simple idea is to code line 5 with respect to line 4 by the differences in number of pixels: $-1, 0, 2, 0, -1$.

Imagine if we send, via a telecommunication channel, the data $-1, 0, 2, 0, -1$ for line 5 after the data for line 4. There would be no confusion nor need to send `7B, 2w, 5B, 2w, 7B` which includes more characters.

Of course, an image can also be scanned column by column, or by a *zig-zag* scan (Figure 10.5).

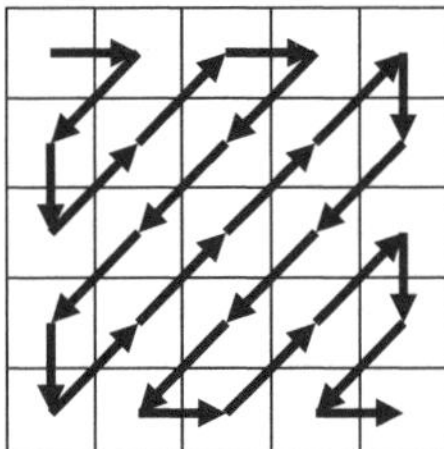

Figure 10.5: A zig-zag scan

Note: the algorithm assumes that successive lines have the same number of runs. The actual ITU-T T.4 algorithm is much more complicated than this.

In practice, to avoid preprocessing, a statistical estimation is applied. The compression model used here is the so-called *Capon model* (Figure 10.6) which was proposed by J. Capon in 1959. A two-state Markov model with states W and B is used. The transmission probabilities $p(W|B)$ and $p(B|W)$ are considered as well as the $p(W)$ and $p(B)$, where $p(W|B)$ is the probability of switching from state B to W, and $p(B|W)$ from W to B; $p(W)$ is the probability of being in state W and $p(B)$ in state B. For facsimile images, $p(W|W)$ and $p(W|B)$ are significantly higher than $p(B|B)$ and $p(B|W)$.

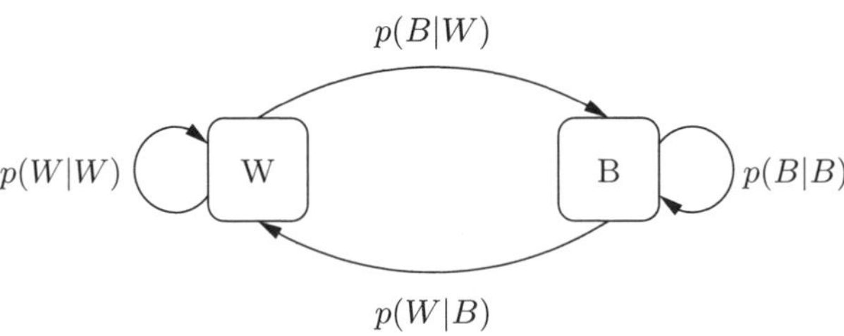

Figure 10.6: The Capon model for binary images

Note: the model is of second order so it expects a better estimation than the first-order model.

10.8.1.2 Extended approach

The idea of this approach extends the image compression principles and concludes that if the current pixel has colour B (or W) then black (or white) pixels seen in the past (or those that will be found in future) tend to have the same immediate neighbours.

The approach checks n of the near neighbours of the current pixel and assigns the neighbours an n bit number. This number is called the *context* of the pixel. In principle there can be 2^n contexts, but the expected distribution is non-uniform because of the image's redundancy.

We then derive a probability distribution of the contexts by counting the occurrence of each context. For each pixel, the encoder can then use adaptive arithmetic coding to encode the pixel with the probabilities. This approach is actually used by JPEG.

10.8.2 Greyscale and colour images

The Graphics Interchange Format (GIF) was introduced by CompuServe in 1987.

In GIF, each pixel of the image is an index to a table that specifies a colour map for the entire image. There are only 256 different colours in the whole image. Of course, the colours may also be chosen from a predefined and much larger palette. GIF allows the colour table to be specified for each image, or for a group of images sharing the use of a map or without a map at all.

10.8.2.1 Reflected Gray codes (RGC)

The Reflected Gray codes (RGC) are a good representation for coding the colours of greyscale images. In this system, we assign codewords in such a way that any two consecutive numbers have codewords differing by 1 bit only.

Example 10.5 *We show below the decimal numbers that are represented by 1 bit, 2 bit and 3 bit RGC accordingly:*

decimal value	*0*	*1*	*2*	*3*	*4*	*5*	*6*	*7*
1 bit RGC	*0*	*1*						
2 bit RGC	*00*	*01*	*11*	*10*				
3 bit RGC	*000*	*001*	*011*	*010*	*110*	*111*	*101*	*100*

An RGC codeword can be derived from a normal binary codeword as follows:

Given a decimal number m, its RGC codeword is
m_2 XOR *shift-1-bit-to-right*(m_2),
where m_2 represents the binary codeword of m.

Example 10.6 *Derive a 3 bit reflected Gray codeword for decimal 3.*

Solution The 3-bit binary code of 3 is: `011`

Shift `011` one bit to the right (and add 0 in front): `001`

`011 XOR 001 = 010`

So the 3-bit RGC codeword for 3 is 010.

10.8.2.2 Dividing a greyscale image

Using the RGC, this approach separates a greyscale image into a number (say n) of bi-level images and applies to each bi-level image a different compression algorithm depending on the characteristics of each bi-level image.

The idea is to assume intuitively that two *similar* adjacent pixels in the greyscale image are likely to be identical in most of the n bi-level images. By 'similar' in the examples below, we mean that the number of different bits between two codewords, i.e. the Hamming distance of two binary codewords, is small as well as the difference in value. For example, 0000 and 0001 are similar because their value difference is 1 and the number of different bit(s) is also 1. The two codewords are identical in the first three bits.

Now we look at an example of separating a greyscale image into n bi-level images.

Example 10.7 *Given a greyscale image with eight shades of grey, we can represent each shade by 3 bits. Let each of the 3 bits, from left to right, be identified as the high, middle and low (bit).*

Suppose that part of a greyscale image is described by matrix `A` *below, where each RGC codeword represents a pixel with the shade of that value:*

```
    010  010  011  110
A = 001  011  010  111
    000  001  011  101
```

Then the image can be separated into three bi-level images (also called bitplanes) as follows:

1. *Bitplane* `A.high` *below consists of all the high bits of* `A`. `A.high` *can be obtained from* `A` *by removing, for each entry of* `A`*, the two bits other than the* `high` *bit.*

```
         0  0  0  1
A.high = 0  0  0  1
         0  0  0  1
```

2. `A.middle` *below consists of all the middle bits of* `A`. `A.middle` *can be obtained from* `A` *by removing, for each entry of* `A`*, the two bits other than the* `middle` *bit.*

```
           1  1  1  1
A.middle = 0  1  1  1
           0  0  1  0
```

3. `A.low` *below consists of all the low bits of* `A`. `A.low` *can be obtained from* `A` *by removing, for each entry of* `A`*, the two bits other than the* `low` *bit.*

```
          0  0  1  0
A.low  =  1  1  0  1
          0  1  1  1
```

As we can see, from bitplane **A.high** to bitplane **A.low**, there is more and more alteration between 0 and 1. The bitplane **A.low** in this example features more random distribution of 0s and 1s, meaning less correlation among the pixels.

To achieve an effective compression, different compression methods should be applied to the bitplanes. For example, we could apply the run-length method to *A.high*, and *A.middle*, and the Huffman coding to *A.low*.

10.8.2.3 JPEG lossless coding

The predictive encoding (Section 8.1.1) is sometimes called *JPEG lossless coding* because the JPEG standard (codified as the ITU-T T.2 Recommendation) specifies a similar predictive lossless algorithm.

The algorithm works with seven possible predictors, i.e. the prediction schemes. The compression algorithm chooses a predictor which maximises the compression ratio. Given a pixel pattern (see below), the algorithm predicts the pixel 'x' in one of eight ways:

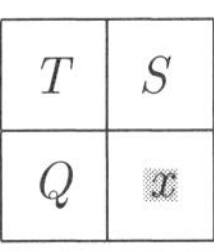

1. No prediction
2. $x = Q$
3. $x = S$
4. $x = T$
5. $x = Q + S - T$
6. $x = Q + (S - T)/2$
7. $x = S + (Q - T)/2$
8. $x = (Q + S)/2$.

The lossless JPEG[2] algorithm gives approximately 2:1 compression ratios on typical greyscale and colour images, which is well superior to GIF (although GIF may be competitive on icons or the like).

[2]It is generally for lossy standard but does have the lossless version and is an extensive and complicated standard for common use.

10.8.3 Lossy compression

Lossy compression aims at achieving a good compression ratio, but the cost for it is the loss of some original information. How much source information can be lost depends very much on the nature of applications. Nevertheless, the measure of a lossy compression algorithm or a lossy compression system normally includes a measure of the quality of reconstructed images compared with the original ones.

10.8.3.1 Distortion measure

Although the best measure of the closeness or fidelity of a reconstructed image is to ask a person familiar with the work to look at the image and provide an opinion; this is not always practical because it is not useful in mathematical design approaches.

Here we introduce the more usual approach in which we try to measure the difference between the reconstructed image and the original one.

There are mathematical tools that measure the distortion in value of two variables, also called *difference distortion measure*. Considering an image to be a matrix of values; the measure of lossy compression algorithm normally uses a standard matrix to measure the difference between reconstructed images and the original ones.

Let P_i be the pixels of reconstructed image and Q_i be the ones of the original, where $i = 1, \cdots, N$. We have the following commonly used measures:

- Squared error measure matrix (this is a measure of the difference):

$$D_i = (P_i - Q_i)^2$$

- Absolute difference measure matrix:

$$D_i = |P_i - Q_i|$$

- Mean squared error measure (MSE) matrix (this is an average measure):

$$\sigma^2 = \frac{1}{N}\sum_{i=1}^{N}(P_i - Q_i)^2$$

- Signal-to-noise-ratio (SNR) matrix (this is the ratio of the average squared value of the source output σ_x^2 and the MSE σ_d^2):

$$\text{SNR} = \frac{\sigma_x^2}{\sigma_d^2}$$

This is often in logarithmic scale (dB):

$$\text{SNR} = 10\log_{10}\frac{\sigma_x^2}{\sigma_d^2}$$

- Peak-signal-to-noise-ratio (PSNR) matrix (this measures the error relative to the average squared value of the signal, again normally in logarithmic scale (dB)):

$$\mathrm{SNR} = 20 \log_{10} \frac{\max_i |P_i|}{\sigma_d^2}$$

- Average of the absolute difference matrix:

$$D_i = \frac{1}{N} \sum_i^N |P_i - Q_i|$$

- Maximum value of the error magnitude matrix:

$$D_{i\infty} = \max_n D_i = |P_i - Q_i|$$

10.8.3.2 Progressive image compression

The main idea of progressive image compression is to gradually compress an image following a underlined order of priority. For example, compress the most important image information first, then compress the next most important information and append it to the compressed file, and so on.

This is an attractive choice when the compressed images are transmitted over a communication channel, and are decompressed and viewed in real time. The receiver would be able to view a development process of the image on the screen from a low to a high quality. The person can usually recognise most of the image features on completion of only 5–10% of the decompression.

The main advantages of progressive image compression are that:

1. The user can control the amount of loss by means of telling the encoder when to stop the encoding process.
2. The viewer can stop the decoding process early since she or he can recognise the image's feature at an early stage.
3. As the compressed file has to be decompressed several times and displayed with different resolution, the decoder can, in each case, stop the decompression process once the device resolution has been reached.

There are several ways to implement the idea of progressive image compression:

- Using so-called SNR progressive or quality progressive compression (i.e. encode spatial frequency data progressively).
- Compress the grey image first and then add the colour. Such a method normally features slow encoding and fast decoding.
- Encode the image in layers. Early layers are large low-resolution pixels followed by smaller high-resolution pixels. The progressive compression done in this way is also called *pyramid coding* or *hierarchical coding*.

Example 10.8 *The following methods are often used in JPEG:*

- *Sequential coding (baseline encoding): this is a way to send data units following a left-to-right, top-to-bottom fashion.*
- *Progressive encoding: this transmits every* n*th line before filling the data in the middle.*
- *Hierarchical encoding: this is a way to compress the image at several different resolutions.*

10.8.4 JPEG (still) image compression standard

This is one of the most widely recognised standards in existence today. It provides a good example of how various techniques can be combined to produce fairly dramatic compression results. The baseline JPEG compression method has a wide variety of hardware and software implementations available for many applications.

JPEG has several lossy encoding modes, from so-called *baseline sequential mode* to *lossless encoding mode*.

The basic steps of lossy JPEG algorithm include processing 24 (or 32) bit colour images and offering a trade-off between compression ratio and quality.

10.8.4.1 Transforms

An image can be compressed if its *correlated* pixels are transformed to a new representation where the pixels become less correlated, i.e. are *decorrelated*. Compression is successful if the new values are smaller than the original ones on average. Lossy compression can be achieved by quantisation of the transformed values. The decoder normally reconstructs the original data from the compressed file applying the opposite transform.

There are a lot of techniques and algorithms for image data compression. You may consult various sources in literature. Due to the limitations of space, we have only provided a sample of the techniques here.

10.8.5 Image file formats

There is lot of image software available these days. They have been developed on various platforms and used different formats or standards. Today's most widely used image formats include:

- GIF (Graphics Interchange Format)
- JPEG (Joint Photographic Experts Group standard for compressing still images).[3]
- Animated GIF (Animated Graphics Interchange Format)

[3] We often use JPEG to mean the standard by this expert group, rather than the organization itself.

- BMP (Windows Bitmap format)
- XBM (X window Bitmap format)
- EPS (Encapsulated PostScript file format)
- PNG (Portable Network Graphics)
- PSD (Adobe Photoshop's native file format)
- PSD Layered (Adobe Photoshop's native file format)
- STN (MediaBin's proprietary STiNG file format)
- TIFF (Tagged Image File Format)
- TGA (TARGA bitmap graphics file format).

Most of these image file formats are in fact compressed, for example, JPEG, PNG, BMP and GIF.

JPEG offers great mainly lossy compression and is widely used for bitmap images. It supports progression and hierarchical mode, and Huffman coding.

PNG supports up to 48 bits per pixel for colour images. It applies LZW compression algorithms and is widely used for the Internet application.

BMP is the native image format in the Microsoft Windows operating systems. Colour depth can be 1, 4, 8, 16, 24 or 32 bits. BMP supports simple run-length compression for 4 and 8 bits per pixel.

GIF was the first universally accepted image format but ended due to legal problems. LZW compression methods support GIF.

Summary

Digital images can be classified as two types: bitmaps and vector graphics. Two commonly used colour representation systems are based on the RGB model and LC representation. Commonly used image files are in formats such as GIF, JPEG, EPS, PNG, PSD, PSD layered, STN, TIFF and TGA. Using simple bit planes to represent digital images, we can apply different compression algorithms that we have learnt so far to get a flavour of image compression systems.

Learning outcomes

On completion of your studies in this chapter, you should be able to:

- illustrate how a picture can be represented by either a bitmap image or a vector image
- explain the main differences between bitmap and vector graphics in terms of visual characteristics

- illustrate RGB colour model and LC representation
- describe a few of the commonly used image formats
- provide examples of lossless and lossy image compression techniques
- illustrate how to represent a greyscale image by several bi-level images
- describe the main ideas of predictive encoding
- be familiar with various distortion measurements
- explain the general approaches of various popular lossy image compression techniques such as progressive image compression and transforms.

Exercises

E10.1 If you consider the three components of an RGB colour to be Cartesian coordinates in a three-dimensional space, and normalise them to lie between 0 and 1, you visualise RGB colour space as a unit cube, see Figure 10.7.

(a) What colours correspond to the eight corners of this cube?

(b) What does the straight line running from the origin to (1, 1, 1) shown in the figure represent?

(c) Comment on the usefulness of this representation as a means of visualising colour.

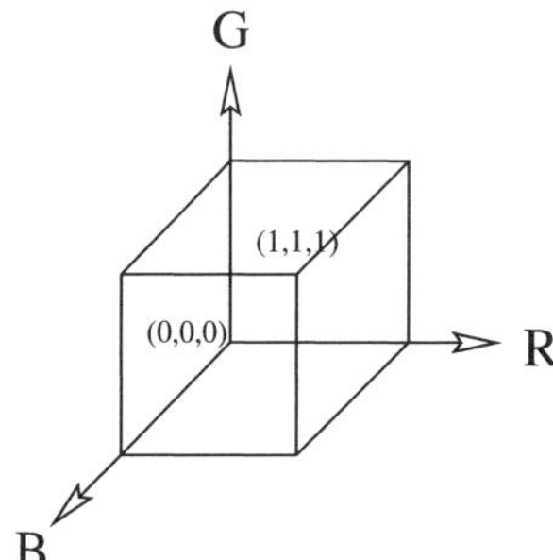

Figure 10.7: The RGB colour space

E10.2 Classify and identify the following images for a continuous-tone image, discrete-tone image and cartoon-like image:

- a reproduction of your national flag
- a photograph of the surface of Mars
- a photograph of yourself on the beach
- a still from an old black-and-white movie.

E10.3 Explain how RGB colour values, when $R = G = B$, can be used to represent shades of grey.

E10.4 Construct a reflected Gray code (RGC) for decimal numbers $0, 1, \cdots, 15$.

E10.5 Following the above question, provide an example to show how a greyscale image with 16 shades of grey can be separated into four bi-level images.

E10.6 Suppose that the matrix **A** below represents the pixel values of part of a large greyscale image, where $i = 0, \cdots, 7$, $j = 0, \cdots, 7$, and $A[0,0] = 4$. Apply any three of the predictive rules of JPEG (below) to matrix **A**.

(a) no prediction[4]
(b) $x = \mathbf{Q}$
(c) $x = \mathbf{S}$
(d) $x = \mathbf{T}$
(e) $x = \mathbf{Q} + \mathbf{S} - \mathbf{T}$
(f) $x = \mathbf{Q} + (\mathbf{S} - \mathbf{T})/2$
(g) $x = \mathbf{S} + (\mathbf{Q} - \mathbf{T})/2$
(h) $x = (\mathbf{Q} + \mathbf{S})/2$.

$$\mathbf{A} = \begin{matrix} 4 & 8 & 4 & 8 & 1 & 1 & 1 & 1 \\ 1 & 2 & 4 & 6 & 5 & 1 & 1 & 1 \\ 8 & 4 & 5 & 5 & 5 & 5 & 5 & 5 \\ 2 & 4 & 8 & 5 & 7 & 9 & 5 & 5 \\ 2 & 4 & 6 & 7 & 7 & 7 & 9 & 9 \\ 2 & 2 & 2 & 3 & 4 & 9 & 7 & 3 \\ 3 & 3 & 6 & 6 & 6 & 7 & 7 & 7 \\ 7 & 7 & 7 & 7 & 6 & 7 & 8 & 8 \end{matrix}$$

E10.7 Implement the ideas above using a higher level programming language.

Laboratory

L10.1 Investigate the colour utilities (or system, package, software) on your computer.

For example, if you have access to Linux, you can go to `Start` → `Graphics` → `More Graphics` → `KColorChooser` and play with the colour display.

L10.2 Experiment on how the values of (R, G, B) (or (H, S, V)) affect the colour displayed.

L10.3 Design and implement a program method which takes a binary codeword and returns its RGC codeword.

[4]i.e. using the original pixel value matrix (**A** in Example 8.2).

L10.4 Design and implement a program method which takes a matrix of integers and returns its residual matrix, assuming that each entry follows JPEG rule (3).

L10.5 Investigate what image formats are supported by a painting or drawing software on your computer. Do they offer any compression?

Assessment

S10.1 Illustrate the RGB colour space using Cartesian coordinates in a three-dimensional space.

S10.2 Consider a square image to be displayed on screen. Suppose the image dimension is 20×20 square inches. The device resolution is 800 dpi. What would be the pixel dimension required?

S10.3 Derive the RGC codeword for a decimal 5.

S10.4 Explain, with an example, how to represent part of the greyscale image below by three bitplanes:

```
000 001 011 011
001 001 001 010
011 001 010 000
```

Bibliography

[Mia99] J. Miano. *Compressed Image File Formats.* Addison Wesley Longman Inc., 1999.

[PM93] W.B. Pennebaker and J.L. Mitchell. *JPEG Still Image Data Compression Standard.* Van Nostrand Reinhold, New York, 1993.

[TM02] D.S. Taubman and M.W. Marcellin. *JPEG2000: Image Compression Fundamentals, Standards and Practice.* Kluwer Academic Publishers, 2nd printing edition, 2002.

[TZ94] D. Taubman and A. Zakhor. Multirate 3D subband coding with motion compensation. *IEEE Transactions on Image Processing*, IP-3:572–588, September 1994.

Chapter 11

Video compression

Video is continuous media in which information is presented to give an illusion of continuity. Sound, movies and computer animation are common examples of continuous media. Temporal characteristics are the focus of the compression potential for continuous media.

Video can be viewed as a sequence of moving pictures played with its audio accompaniment. The moving pictures and audio signals, as two components, may be dealt with separately. Moving pictures consist of a time series of still images called *frames*. To feel alive, each frame of image can be perceived only in the existence of its previous and succeeding frames.

Video files contain a mixture of almost everything: text, graphics, audio, images and animation data. Video systems change the images 20 to 30 times per second. So the amount of data involved is enormous. Video compression algorithms are lossy, making use of human visual and audio perception systems' forgiving nature.

Video systems can be classified broadly into two types: old *analogue video* and modern *digital video*.

11.1 Analogue video

The essential function of analogue video is to display still pictures on a television receiver or a monitor screen one frame after another. The pictures are converted into an electronic signal by a raster scanning process. Conceptually, a screen is divided into horizontal lines. The picture we see is built up as a sequence of horizontal lines from top to bottom. Here the human persistence of vision plays an important part and makes the series of lines appear as a (frame of) picture.

The following parameters determine the quality of pictures:

- number of scanning lines
- number of pixels per scan line

- number of displayed frames per second (in *fps*): this is called the *frame rate*
- scanning techniques.

For example, the screen has to be refreshed about 40 times a second, otherwise, it will flicker. Popular standards include Phase Alternating Line (PAL), Sequential Couleur Avec Memoire (SECAM) and National Television Systems Committee (NTSC). PAL is mainly used in most of Western Europe and Australia, China and New Zealand. SECAM is used in France, in the former Soviet Union and in Eastern Europe. NTSC is used in North America, Japan, Taiwan, part of the Caribbean, and in South America.

A PAL or SECAM frame contains 625 lines, of which 576 are for pictures. An NTSC frame contains 525 lines, of which 480 are pictures. PAL or SECAM uses a frame rate of 25 fps and NTSC 30 fps.

11.2 Digital video

Although the idea of digital video is simply to sample the analogue signal and convert it into a digital form, the standard situation is inevitably quite complex due to the need to compromise the existing equipment in old as well as new standards such as HDTV.

The sample standard is ITU-R BT.601 (more commonly known as CCIR 601), which defines sampling of digital video. It specifies a horizontal sampling picture format consisting of 720 luminance samples and two sets of 360 colour difference samples per line. The size of PAL screen frames is 768×576 and NTSC 640×480.

Apart from the digital data from analogue video sampling, digital video data to be compressed also include the data streams produced by various video equipment such as digital video cameras and VTRs. Once the data are input into the computer there are many ways to process them.

11.3 Moving pictures

Video data can be considered as a *sequence* of still images changing with time. Owing to the forgiving nature of the human eye, video signals change the image 20 to 30 times per second. Video compression algorithms all tend to be lossy.

One minute of modern video may consist of 1500–1800 still images. This explains the need and motivation for video compression.

Among the many different video compression standards, there are two important ones, namely *ITU-T H.261* and *MPEG*. *ITU-T H.261* is intended for use mainly for videoconferencing and videotelephony, and *MPEG* is mainly for computer applications.

11.4 MPEG

- This is the standard designed by the Motion Picture Expert Group, including the updated versions MPEG-1, MPEG-2, MPEG-4, and MPEG-7.
- Most popular standards include the ISO's MPEG-1 and the ITU's H.261.
- It is built upon the three basic common analogue television standards: NTSC, PAL and SECAM.
- There are two sizes of SIF (Source Input Format): SIF-525 (with NTSC video) and SIF-625 (for PAL video).
- The H.261 standard uses CIF (Common Intermediate Format) and QCIF (Quarter Common Intermediate Format).

As we can see, MPEG is greatly influenced by ITU-T standards and even includes some of its standards. This may be the reason why ITU-T H.261 is the more fully developed of the two.

11.5 Basic principles

Video compression is based on two types of redundancies among the video data, namely *spatial redundancy* and *temporal redundancy*.

1. Spatial redundancy means the correlation among neighbouring pixels in each frame of image. This can be dealt with by the techniques for compressing still images (Chapter 10).
2. Temporal redundancy means the similarity among neighbouring frames, since a video frame tends to be similar to its immediate neighbours.

Techniques for removing spatial redundancy are called *spatial compression* or *intra-frame* compression. Techniques for removing temporal redundancy are called *temporal compression* or *inter-frame* compression. We focus on temporal compression here, since for spatial compression many techniques for still images in Chapter 10 can be applied.

11.6 Temporal compression algorithms

In these algorithms, certain frames in a sequence are identified as *key frames*. These key frames are often specified to occur at regular intervals. The key frames are either left uncompressed or are more likely to be spatially compressed. Each of the frames between the key frames is replaced by a so-called *difference frame* which records the differences between the original frames and the most recent key frame. Alternatively, the differences between the original frames and the preceding frame can also be stored in the difference frame depending on the sophistication of the decompressor.

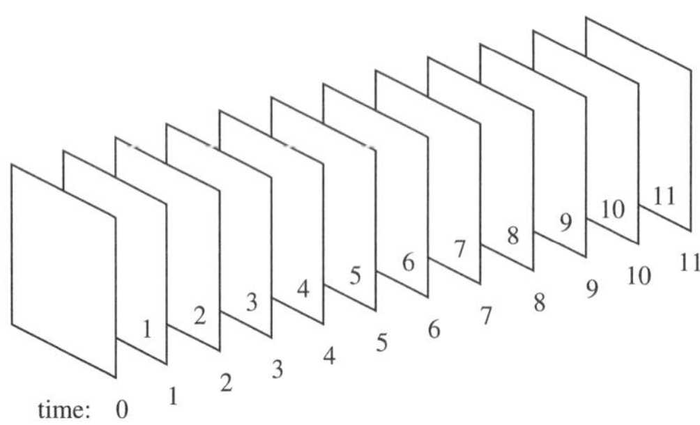

Figure 11.1: A sequence of original frames

Example 11.1 *Figure 11.1 shows a sequence of frames to be displayed during a time duration $0, 1, \cdots, 11$.*

Following the MPEG regulation, the frames marked 'I' (as in Figure 11.2) are encoded as the key frames. These are called *I-pictures*, meaning *intra*, and have been compressed purely by spatial compression. Difference frames compared with previous frames are called *P-pictures* or *predictive pictures*. The frames marked *B* are so-called *B-pictures* which are predicted from later frames.

Example 11.2 *The encoded frames are as in Figure 11.2.*

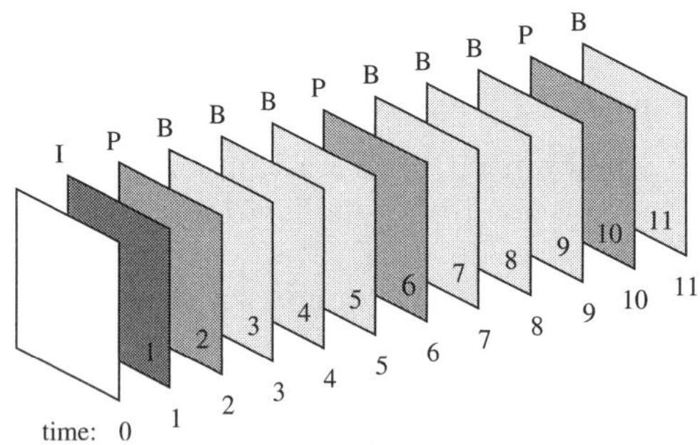

Figure 11.2: Encode frames as I-, P- and B-pictures

The decoding process now becomes a question of dealing with a sequence of compressed frames with I, P, B types of pictures.

I-pictures are decoded independently. P-pictures are decoded using the preceding I or P frames. B-pictures are decoded according to both preceding and following I- or P-pictures. The encoded frames can then be decoded as IBBBPBBBP and are displayed in that order (see Figure 11.4).

Example 11.3 *Note that the order of the encoded and displayed frames in Figure 11.3 is altered slightly from that in Figure 11.2.*

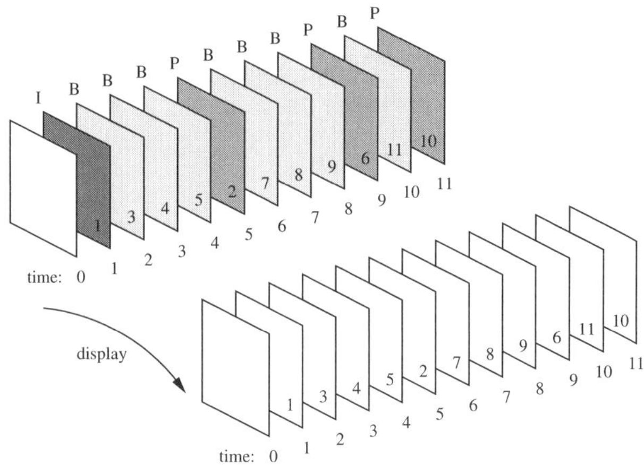

Figure 11.3: Decoded and displayed from I-, P- and B-pictures

11.7 Group of pictures

Now a video clip can be encoded as a sequence of I-, P- and B-pictures. Group of pictures (GOP) is a repeating sequence beginning with an I-picture and is used by encoders.

Example 11.4 *Figure 11.4 shows a GOP sequence IBBPBB containing two groups.*

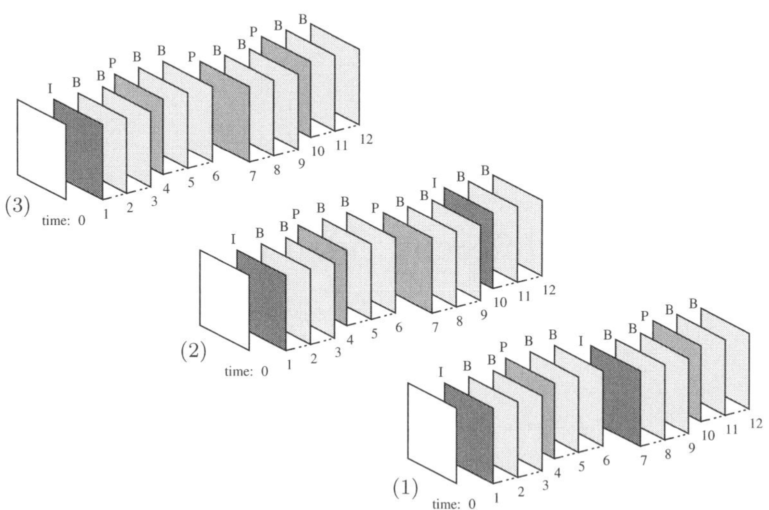

Figure 11.4: GOPs (1) IBBPBB, (2) IBBPBBPBB , (3) IBBPBBPBBPBB

Other common sequences in use are IBBPBBPBB with three groups and IBBPBBPBBPBB with four groups.

11.8 Motion estimation

The main idea is to represent each block of pixels by a vector showing the amount of block that has moved between images. The approach is most successful for compressing a background of images where the fixed scenery can be represented by a fixed vector of (0, 0).

It uses the following methods:

- Forward prediction: computing the difference only from a previous frame that was already decoded.
- Backward prediction: computing the difference only from the next frame that is about to be decoded.
- Bidirectional prediction: the pixels' information in a frame depends on both the last frame just decoded and the frame about to follow.

The following factors can be used to estimate motion:

- Basic difference
- L_1 norm: mean absolute error

$$||A - B||_1 = \sum_i \sum_j |A_{i,j} - B_{i,j}|$$

- L_2 norm: mean squared error

$$||A - B||_2 = \sum_i \sum_j (A_{i,j} - B_{i,j})^2$$

11.9 Work in different video formats

To allow systems to use different video formats, the following process is needed:

- Standardise the format
- Break the data into 8×8 blocks of pixels
- Compare with other blocks
- Choose important coefficients
- Scale the coefficients for quantisation
- Pack the coefficients.

Summary

Video data is a good example of digital multimedia today. The massive amount of data demonstrates clearly the potential of data compression. For continuous media, the compression techniques introduced in the previous chapters can be used to conduct various kinds of spatial compression.

Learning outcomes

On completion of your studies in this chapter, you should be able to:

- identify and list the parameters that determine the quality of pictures
- explain the popular standards for analogue and digital videos
- describe, with an example, how temporal compression algorithms work in principle
- explain the common measures for motion estimation.

Exercises

E11.1 Explain briefly the concepts of spatial redundancy and temporal redundancy.

E11.2 Implement the basic *temporal compression algorithm* introduced in the chapter.

Assessment

S11.1 Explain the two characters of video data.

S11.2 Illustrate how a temporal compression algorithm works.

Bibliography

[HV02] D.T Hoang and J.S. Vitter. *Efficient Algorithms for MPEG Video Compression.* John Wiley & Sons, New York, 2002.

[Sol97] S.J. Solari. *Digital Video and Audio Compression.* McGraw Hill Text, 1997.

[Sym98] P. Symes. *Video Compression.* McGraw-Hill Companies, 1998.

[Sym03] P.D. Symes. *MPEG-4 Demystified.* McGraw-Hill Professional Publishing, January 2003.

Appendix A

Brief history

Data compression

Data compression can be viewed as the art of creating shorthand representations for the data, but this process started as early as 1000 BC. The short list below gives a brief survey of the historical milestones:

- 1000 BC shorthand
- 1829 Braille code
- 1838 Morse code was invented for use in telegraphy
- 1843 variable-length codes for telegraph
- 1930 analog compression
- 1948 information theory
- 1949 coding blocks based on their probabilities
- 1951 Huffman codes
- 1966 Run-length coding
- 1976 arithmetic coding
- 1970s dynamic Huffman coding
- 1977 dictionary-based compression
- 1982 LZSS (by Storer and Szymanski)
- 1984 LZW (LZ was improved by Terry Welch); pulse position modulation (PPM)
- 1987 dynamic Markov compression (DMC)
- 1992 gzip
- 1994 Burrows and Wheeler transform (BWT)
- 1996 bzip2

- 1997 PPM2
- 1980s
  - early 1980s FAX
  - mid-1980s video conferencing, still images (JPEG), improved FAX standard (JBIG)
  - late 1980s onward motion video compression (MPEG)
- 1990s
  - early 1990s disk compression (stacker)
  - mid-1990s satellite TV
  - late 1990s digital TV (HDTV), DVD, MP3
- 2000s digital TV (HDTV), DVD, MP3.

Audio compression

MP3 is more popular than 'audio compression'. Everyone knows MP3 but not everyone knows that MP3 stands for MPEG layer III.

- 1977 idea of creating a method of transferring music over a standard phone line
- 1987 Digital Audio Broadcasting (DAB) in Fraunhofer Institute
- 1988 MPEG was established as a subcommittee of ISO/IEC
- 1989 Fraunhofer received a German patent for MP3
- 1992 Fraunhofer's and Dieter Seitzer's audio coding algorithm was integrated into MPEG1
- 1991 MP3 codec
- 1993 MP2, MPEG1 audio layer II
- 1993 Maplay by Tobias Bading
- 1995 MPEG2 was published
- 1996 US patent issued for MP3
- 1997 AMP the first MP3 playback engine
- 1998 Fraunhofer enforced their patent rights
- 1998, Winamp, a free MP3 music player (by students Justin Frankel and Dmitry Boldyrev at the time)
- 1999 SubPop, a record company, first started to distribute music tracks in the MP3 format
- 1999 portable MP3 players available
- 2001 ID3.

Image compression

- 1990 the first derivatives of many signals exhibited lower information content than the source signal
- 1980 delta modulation and differential pulse code modulation (DPCMs)
- 1977 recursive and adaptive quantisation techniques
- 1995 Block Encoding (BE), Vector Quantisation (VQ) codebook
- 1986, 1993 singular Value Decomposition (SVD) (Karhunen-Loeve transform (KLT))
- 1987, 1994, transform coding such as Fourier Transform (FT) or Cosine Transform (CT)

Video compression

The brief history of computer applications below shows the dramatic progress made in recent years due to the availability of the techniques for computer images and sounds.

- 1940s electronic computers performing numerical applications
- 1950s first non-numerical applications: texts
- 1960s analogue videophone system, still images
- 1970s computer animations
- 1980s digital sound
- 1990s television broadcasters started using MPEG2 coded digital forms
- present multimedia applications, communications and entertainment.

Wavelets

- 1807 Fourier series
- 1909 Haar wavelet
- 1930 Haar basis function and a function that can vary in scale and conserve energy
- 1980 an effective algorithm for numerical image processing using wavelets
- 1960–80 atoms and assembly rules by various researchers
- 1980 Meyer wavelets
- 1982 Ingrid Daubechies's set of wavelet orthonormal basis functions, Marr wavelet.

Appendix B

Matrices

An $m \times n$ matrix (or m by n matrix) is a rectangular array of data consisting of m *rows* and n *columns* and can be represented as follows:

$$\mathbf{A} = (a_{ij})_{m\times n} = \begin{pmatrix} a_{11} & \cdots & a_{1n} \\ \vdots & & \vdots \\ a_{m1} & \cdots & a_{mn} \end{pmatrix}$$

The $m \times n$ is called the *dimension* or *order* of the matrix. Each datum a_{ij} in the matrix $(a_{ij})_{m\times n}$ is called an *entry* of the matrix. Each entry has a *row index* i and *column index* j, together representing the position of the entry in the matrix.

Example B.1 *A matrix of* 4×3

$$\mathbf{A} = \begin{pmatrix} 5 & 2 & 4 \\ 1 & -3 & -1 \\ 1 & -1 & 1 \\ 3 & 2.5 & 2 \end{pmatrix}$$

B.1 Special matrices

1. Square matrix:

   A *square* matrix is a matrix of order $m = n$ (short for $n \times n$).

   Example B.2 *A square matrix of order* 3

$$\mathbf{A} = \begin{pmatrix} 5 & 2 & 4 \\ 1 & -3 & -1 \\ 1 & -1 & 1 \end{pmatrix}$$

2. Vector:

   A *vector* is a matrix of $m \times 1$ or $1 \times n$.

3. Null matrix (or zero matrix):

   A matrix is called a *null matrix* if all the entries are zero, and can be denoted by $0_{m\times n}$, or 0 for short.

   Example B.3

$$\mathbf{0} = \mathbf{0_{3\times 4}} = \begin{pmatrix} 0 & 0 & 0 \\ 0 & 0 & 0 \\ 0 & 0 & 0 \\ 0 & 0 & 0 \end{pmatrix}$$

4. Scalar:

   If $m = n = 1$, the matrix becomes a single entry and is called *scalar*. In this case, the a_{11} is used without the bracket.

5. Identity matrix:

   A matrix is an identity matrix if it is square, and all the data on the diagonal of the matrix are 1, and all the other entries are zero.

   Example B.4

$$\mathbf{I} = \mathbf{I_3} = \begin{pmatrix} 1 & 0 & 0 \\ 0 & 1 & 0 \\ 0 & 0 & 1 \end{pmatrix}$$

6. Inverse:

   Let $\mathbf{A}$ be a square matrix of order n. If there exists another square matrix $\mathbf{B}$ of order n, such that the matrix product $\mathbf{AB} = \mathbf{BA} = \mathbf{I}$, then $\mathbf{A}$ is said to be *invertible* with *inverse* $\mathbf{B}$. Matrix $\mathbf{B}$ can be shown to be unique, and is called the *inverse* of $\mathbf{A}$ and is denoted as $\mathbf{A^{-1}}$.

7. Transpose:

   Let $\mathbf{A}$ be a square matrix (a_{ij}) of order $m \times n$.

$$\mathbf{A} = \begin{pmatrix} 5 & 2 & 4 \\ 1 & -3 & -1 \\ 1 & -1 & 1 \\ 3 & 2.5 & 2 \end{pmatrix}$$

   The transpose $\mathbf{A}^T$ is of order $n \times m$ and becomes (a_{ji}).

$$\mathbf{A}^T = \begin{pmatrix} 5 & 1 & 1 & 3 \\ 2 & -3 & -1 & 2.5 \\ 4 & -1 & 1 & 2 \end{pmatrix}$$

8. Symmetric matrix:

   A square matrix is *symmetric* if $\mathbf{A}^T = \mathbf{A}$. In a symmetric matrix, the elements on one side of the diagonal are mirror images of those on the other side.

Example B.5 *The following matrix ($\boldsymbol{A}_3$) is symmetric:*

$$\mathbf{A}_3 = \begin{pmatrix} 1 & 3 & 2 \\ 3 & 5 & 0 \\ 2 & 0 & 6 \end{pmatrix}$$

B.2 Matrix operations

1. Addition and subtraction:

   Addition and subtraction can only performed on two matrices of the same dimension (or order).

$$(a_{ij})_{m\times n} + (b_{ij})_{m\times n} = (a_{ij} + b_{ij})_{m\times n}$$

$$(a_{ij})_{m\times n} - (b_{ij})_{m\times n} = (a_{ij} - b_{ij})_{m\times n}$$

2. Multiplication:

   (a) Multiplication by a scalar k.

$$k\mathbf{A} = k(a_{ij})_{m\times n} = (ka_{ij})_{m\times n}$$

   (b) Matrix product:

   A matrix $\mathbf{A}_{m\times n}$ of order $m \times n$ can only by multiplied by another matrix $\mathbf{B}_{n\times p}$ of order $n \times p$ in that order. The matrix product $\mathbf{A}_{m\times n}\mathbf{B}_{n\times p}$ is a matrix of order $m \times p$, and each product entry is

$$\sum_{k=1}^{n} a_{ik}b_{kj}, \text{ where } i = 1, \cdots, m, j = 1, \cdots, p.$$

   The two matrices **A** and **B** are said to be **conformable for multiplication** if the number of columns of **A** equals the number of rows of **B**.

   Note, matrix multiplication is not commutative in general, i.e. the product **AB** is *not* equal to **BA**.

   In a product **AB** of the two matrices, **B** is said to be *premultiplied* by **A**. **A** is said to be *postmultiplied* by **B**.

   Example B.6

$$\mathbf{A}_{2\times 3} = \begin{pmatrix} 1 & 3 & 2 \\ 3 & 5 & 0 \end{pmatrix}$$

$$\mathbf{B}_{3\times 1} = \begin{pmatrix} 4 \\ 1 \\ 2 \end{pmatrix}$$

Here **A** *and* **B** *are conformable for multiplication.*

$$\mathbf{A}_{2\times 3}\mathbf{B}_{3\times 1} = \begin{pmatrix} 1\times 4+3\times 1+2\times 2 \\ 3\times 4+5\times 1+0\times 2 \end{pmatrix} = \begin{pmatrix} 11 \\ 17 \end{pmatrix}$$

3. Commutative, distributive and associative rules:

   (a) $\mathbf{A}+\mathbf{B}=\mathbf{B}+\mathbf{A}$
   (b) $\mathbf{A}(\mathbf{B}+\mathbf{C})=\mathbf{AB}+\mathbf{AC}$
   (c) $\mathbf{A}(\mathbf{BC})=(\mathbf{AB})\mathbf{C}$
   (d) $\mathbf{AI}=\mathbf{IA}=\mathbf{A}$
   (e) $\mathbf{A0}=\mathbf{0A}=\mathbf{0}$
   (f) $\mathbf{A}^m\mathbf{A}^n=\mathbf{A}^{m+n}$
   (g) $\mathbf{A}^{m^n}=\mathbf{A}^{mn}$.

B.3 Determinants

Let **A** be a square matrix of order n. The determinant of **A** is a *value* of the sums and products of the entries in **A** following specific rules. It is denoted as an array of order n as below:

$$\det \mathbf{A} = \begin{vmatrix} a_{11} & \cdots & a_{1n} \\ \vdots & \cdots & \vdots \\ a_{n1} & \cdots & a_{nn} \end{vmatrix}$$

The calculation rules transform the patterns of numbers into a single number. The determinant can be of a complex value as well as a real one and depends on the entries of **A**. The value of a determinant of order n is the algebraic sum of $n!$ terms, each being the product of n different entries taken one each from *every* row and column of the determinant.

1. $\det \mathbf{a} - a$
2. $\det \mathbf{A} = a_{i1}(-1)^{i+1}M_{il} + a_{i2}(-1)^{i+2}M_{i2} + \cdots + a_{in}(-1)^{i+n}M_{in}$, where M_{ij} is a so-called *principal minor*, and is a subdeterminant obtained by deleting row i and column j from $\det \mathbf{A}$.

Example B.7

$$\begin{aligned} \det\ \mathbf{A} &= \begin{vmatrix} 2 & 0 & 1 \\ 3 & 1 & 1 \\ 1 & 2 & 1 \end{vmatrix} \\ &= 2\times\begin{vmatrix} 1 & 1 \\ 2 & 1 \end{vmatrix} - 0\times\begin{vmatrix} 3 & 1 \\ 1 & 1 \end{vmatrix} + 1\times\begin{vmatrix} 3 & 1 \\ 1 & 2 \end{vmatrix} \\ &= 2\times(1\times 1-1\times 2)+0+1\times(3\times 2-1\times 1) \\ &= 3 \end{aligned}$$

A determinant value remains unchanged if

1. all rows and all columns are transposed without changing the entry order.

   Example B.8

$$\begin{vmatrix} 2 & 0 & 1 \\ 3 & 1 & 1 \\ 1 & 2 & 1 \end{vmatrix} = \begin{vmatrix} 2 & 3 & 1 \\ 0 & 1 & 2 \\ 1 & 1 & 1 \end{vmatrix} = 3$$

2. the entries of any one row, column, or a multiple of them are added to (or subtracted from) another row, column, or multiple, then the value of the determinant remains the same.

   Example B.9 *We multiply the first row by 2 and add it to the second row. The determinant value remains the same.*

$$\begin{vmatrix} 2 & 0 & 1 \\ 3 & 1 & 1 \\ 1 & 2 & 1 \end{vmatrix} = \begin{vmatrix} 2 & 0 & 1 \\ 3+2\times 2 & 1+0\times 2 & 1+2\times 1 \\ 1 & 2 & 1 \end{vmatrix} = 3$$

3. $\det(\mathbf{AB}) = \det(\mathbf{A})(\mathbf{B})$.

Eigenvalues and eigenvectors

The number λ, complex or real, is an *eigenvalue* of the square matrix $\mathbf{A}$ if there is a vector $\mathbf{x} \neq \mathbf{0}$, such that

$$\mathbf{Ax} = \lambda\mathbf{x}$$

The vector $\mathbf{x}$ is called an *eigenvector* and corresponds to the eigenvalue λ. λ is an eigenvalue of $\mathbf{A}$ if and only if

$$\det(\mathbf{A} - \lambda\mathbf{I}) = 0$$

$\det(\mathbf{A} - \lambda\mathbf{I}) = 0$ is called the *characteristic equation* for the square matrix $\mathbf{A}$ and the eigenvalues are the roots of the polynomial of degree n. The polynomial is called *characteristic polynomial* of $\mathbf{A}$.

B.4 Orthogonal matrix

A matrix $\mathbf{A}$ that is equal to the inverse of its transpose matrix, $(\mathbf{A}^T)^{-1}$, is an *orthogonal matrix.*

Example B.10

$$\mathbf{A} = \begin{pmatrix} \frac{1}{\sqrt{2}} & -\frac{1}{\sqrt{2}} \\ \frac{1}{\sqrt{2}} & \frac{1}{\sqrt{2}} \end{pmatrix} \textit{ is orthogonal, and}$$

$$\mathbf{AA}^T = \begin{pmatrix} \frac{1}{\sqrt{2}} & -\frac{1}{\sqrt{2}} \\ \frac{1}{\sqrt{2}} & \frac{1}{\sqrt{2}} \end{pmatrix} \begin{pmatrix} \frac{1}{\sqrt{2}} & \frac{1}{\sqrt{2}} \\ -\frac{1}{\sqrt{2}} & \frac{1}{\sqrt{2}} \end{pmatrix} = \begin{pmatrix} 1 & 0 \\ 0 & 1 \end{pmatrix} \textit{ is orthogonal}$$

If two matrices **A** and **B** are orthogonal and conformable for multiplication, their product **AB** is also an orthogonal matrix. Let **A** be an orthogonal matrix. Then

1. $\mathbf{A}^T\mathbf{A} = \mathbf{AA}^T = \mathbf{I}$
2. $\mathbf{A}^T = \mathbf{A}^{-1}$
3. $\mathbf{AI} = \mathbf{IA} = \mathbf{I}$
4. $\mathbf{A}^T = \mathbf{A}^{-1}$
5. $(\mathbf{AB})^T = \mathbf{B}^T\mathbf{A}^T$
6. $(\mathbf{AB})^{-1} = \mathbf{B}^{-1}\mathbf{A}^{-1}$
7. $(\mathbf{A}^{-1})^T = (\mathbf{A}^T)^{-1}$
8. $\det \mathbf{I} = 1$
9. $\det \mathbf{AB} = \det \mathbf{A} \ \det \mathbf{B}$
10. $\det \mathbf{A}^T = \det \mathbf{A}$.

B.4.1 Inner product

Let **x** and **y** be two vectors of order n. An inner product, also called the *dot product*, between **x** and **y** is defined as

$$\mathbf{x} \cdot \mathbf{y} = x_1 y_1 + x_2 y_2 + \cdots, x_n y_n$$

B.4.1.1 Orthogonal vector

x and **y** are said to be *orthogonal* to each other if their inner product is zero.

B.4.1.2 Orthogonal set

A set of vectors $\mathbf{x_1}, \mathbf{x_2}, \cdots, \mathbf{x_m}$ of order m is said to be orthogonal if each vector is orthogonal to every other vector in the set.

The coefficient of a vector **v** corresponding to a unit vector **u** from an *orthogonal basis set* can be obtained by computing the inner product between the vector and the unit vector. For example, given an orthogonal $\mathbf{u_x} = (1, 0)^T$ and $\mathbf{u_y} = (0, 1)^T$, $\mathbf{v} \cdot \mathbf{u_x} = v_1 \times 1 + v_2 \times 0 = v_1$, and $\mathbf{v} \cdot \mathbf{u_y} = v_1 \times 0 + v_2 \times 1 = v_2$.

B.4.2 Vector space

A vector space consists of a set of vectors with the operations of vector addition and scalar multiplication defined on them. The results of these operations are also elements of the vector space.

Inner product $\mathbf{x} \cdot \mathbf{x} \geq 0$ with $\mathbf{x} \cdot \mathbf{x} = 0$ if and only if $\mathbf{x} = \mathbf{0}$. Then the quantity $\sqrt{\mathbf{x} \cdot \mathbf{x}}$ denoted by $\|\mathbf{x}\|$ is called the *norm* of $\mathbf{x}$ and agrees with our usual concept of Euclidean distance in two- or three-dimensional situations.

Appendix C

Fourier series and harmonic analysis

An important advantage of the Fourier series representation of a function is that it can represent a periodic function containing a number of finite discontinuities with no requisite of the use of successive differential coefficients as in Taylor series.

C.1 Fourier series

In mathematical expression, we have

$$f(t) = \frac{a_0}{2} + \sum_{i=1}^{\infty} a_i \cos(i\frac{2\pi}{T}t) + \sum_{i=1}^{\infty} a_i \sin(i\frac{2\pi}{T}t), \tag{C.1}$$

where $f(t) = f(t + nT)$ and n is an integer.

Using Euler's identity $e^{jx} = \cos\ x + j\sin\ x$, the formula in C.1 can be represented in exponential form.

$$f(t) = \sum_{-\infty}^{\infty} c_n e^{j\frac{2\pi nt}{T}}$$

DFT may be useful for data compression because a few coefficients of the Fourier expansion may be sufficient to make the reconstructed wave close enough to the original function. For example, Figure C.1 shows how only a few terms of the Fourier expression can give quite a good approximation of the original signal. (1) is the original function

$$f(t) = \begin{cases} 1, & 0 \le t < \pi \\ 0, & -\pi \le t < 0 \\ f(t + 2k\pi), & k = 1, 2, \cdots \end{cases}$$

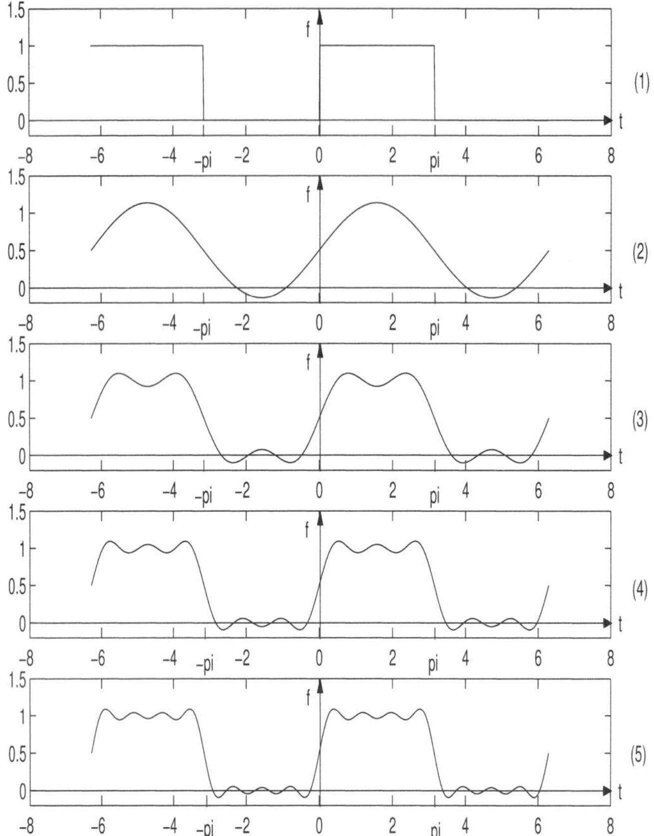

Figure C.1: (1) f(t), (2)–(5) a few Fourier terms

(2) is the approximation with the first harmonic $f(t) \approx \frac{1}{2} + \frac{2\sin t}{\pi}$. (3) is the approximation with the first two harmonic terms. $f(t) \approx \frac{1}{2} + \frac{2}{\pi}(\frac{\sin t}{1} + \frac{\sin 3t}{3})$. (4) is the approximation $f(t) \approx \frac{1}{2} + \frac{2}{\pi}(\frac{\sin t}{1} + \frac{\sin 3t}{3} + \frac{\sin 5t}{5})$. (5) is the approximation $f(t) \approx \frac{1}{2} + \frac{2}{\pi}(\frac{\sin t}{1} + \frac{\sin 3t}{3} + \frac{\sin 5t}{5} + \frac{\sin 7t}{7})$.

Of course, reconstructing precisely the exact original function requires an infinite number of terms (or coefficients) which is impossible in practice anyway.

C.2 Convergent series

$$e^x = 1 + \frac{x}{1!} + \frac{x^2}{2!} + \frac{x^3}{3!} + \cdots \tag{C.2}$$

$$\cos x = 1 - \frac{x^2}{2!} + \frac{x^4}{4!} - \cdots \tag{C.3}$$

$$\sin x = \frac{x}{1!} + \frac{x^3}{3!} + \frac{x^5}{5!} - \cdots \tag{C.4}$$

C.2.1 Euler's identity

$$e^{jx} = \cos x + j \sin x \tag{C.5}$$

or

$$e^{-jx} = \cos x - j \sin x \tag{C.6}$$

$$\cos x = \frac{e^{jx} + e^{-jx}}{2} \tag{C.7}$$

$$\sin x = \frac{e^{jx} - e^{-jx}}{2} \tag{C.8}$$

Appendix D

Pseudocode notation

Pseudocode is merely a convenient way to describe and convey algorithmic ideas. It would not be a big problem if you just use any ad hoc language as long as the algorithmic idea is clear. However, one useful function of pseudocode is that it can serve as a bridge or a translator to convert an algorithm to a source code of a computer program. So an ideal pseudocode would be *close* enough in syntax to a conventional computer language.

In this book, we use a hybrid of adopted keywords and syntax from several commonly used high-level sequential computer languages. Of course, the user can extend the pseudocode vocabulary by adding more useful terms, function names etc.

D.1 Values

- $0, 1, \cdots, 9$
- fraction, rational numbers
- *true*, *false*.

D.2 Types

boolean, int, real, char, string, object.

D.3 Operations

- $\leftarrow$ (assignment)
- (), [], {}
- $+, -, \times, /$
- and (&&), or (||), xor (`XOR`)
- $<, >, \leq, \geq, =, \neq$.

D.4 Priority

In some cases, there are nested structures, or a long expression consists of many items.

- nested brackets: from inside out
- an expression with equal priority: from left to right
- algorithm without line numbers: from the top line down.

D.5 Data structures

- array, list, queue, stack, set
- tree, graph, matrix
- hash-table.

D.6 Other reserved words

- function
- procedure
- method
- (typed) method.

Note: the terms `procedure` and `function` are used in language such as C, C++ and `method` is the concept in Java. In our pseudocode, keywords `function` and `typed method`, and `procedure` and `(void) method` are interchangeable.

D.7 Control keywords

These are the keywords to indicate a block of statements of an algorithm.

- **if** – **end if**
- **for** – **end for**
- **repeat** – **until**
- **while** – **end while**
- **return**.

D.8 Examples of sequential structures

1. A method, function or procedure

   method xxx (type a, b, c)
   input: type a, b, c
   output: $a + b + c$
   other statements

   type **method** xxx (type a, b, c)
   return $a + b + c$

   type **function** xxx (type a, b, c)
   return $a + b + c$

   procedure xxx (type a, b, c)
   return $a + b + c$

2. If-then-else

   if condition **then**
     other statements
   end if

   if condition **then**
     other statements
   else
     other statements
   end if

   if condition **then**
     other statements
   else if condition **then**
     other statements
   else
     other statements
   end if

3. A Boolean function (or method)

   boolean function xxx (int a, b)
   return $a > b$

4. A loop (iteration) structure

   for $i = 1$ to n **do**
   other statements
   end for

   while condition **do**
   other statements
   end while

   repeat
   other statements
   until condition

Appendix E

Notation

The following notation is used in the book:

symbolic data `a`, `b`, ⋯, `z`, `A`, `B`, ⋯, `Z`, `'a'`, `'b'`, ⋯, `'z'`, `'A'`, `'B'`, ⋯, `'Z'`, *`a`*, *`b`*, ⋯, *`z`*, *`A`*, *`B`*, ⋯, *`Z`*, *`'a'`*, *`'b'`*, ⋯, *`'z'`*, *`'A'`*, *`'B'`*, ⋯, *`'Z'`*, $a, b, \cdots, z, A, B, \cdots, Z$, '$a$', '$b$', ⋯, '$z$', '$A$', '$B$', ⋯, '$Z$'

binary data 0, 1, *0*, *1*, a, b, *a*, *b*, b, w, *b*, *w*

lists or sets A, B, ⋯, Z, **A, B, ⋯, Z**, $\mathcal{A}, \mathcal{B}, \cdots, \mathcal{Z}$

functions I(), K(), H(), *I()*, *K()*, *H()*

variables s_j, p_j, c_j, $j = 1, \cdots, n$, $i = 1, \cdots, n$, x, y, s_j, p_j, c_j, $j = 1, \cdots, n$, $i = 1, \cdots, n$, x, y, $a, b, \cdots, z, A, B, \cdots, Z$

matrices $\mathbf{x}, \mathbf{y}$, $\mathbf{A}, \mathbf{B}, \mathbf{C}$, (a_{ij}), (b_{ij}), (c_{ij})

average lengths $\bar{l}$, $\bar{L}$

mathematical symbols $\log x$, $\log_2 x$, $\sin x$, $\cos x$

special symbol Self-information: I(), $I()$; Entropy: H(), $H()$

methods procedures or function: *next_symbol_in_text()*, *update_tree(T)*, *next_symbol_in_text()*, *update_tree(T)*, etc.

Index

CARDIFF UNIVERSITY ★ PRIFYSGOL CAERDYDD